THE DARK MATTER

Also by Wallace Tucker

Radiation Processes in Astrophysics
(MIT Press, 1975)

Heath Physical Science
(with L. Nolan)
(D.C. Heath and Co., 1984)

The Star Splitters
(U.S. Government Printing Office, 1984)

The X-Ray Universe
(with R. Giacconi)
(Harvard University Press, 1985)

Also by Wallace Tucker and Karen Tucker

The Cosmic Inquirers
(Harvard University Press, 1986)

THE DARK MATTER

Contemporary Science's Quest
for the Mass Hidden
in Our Universe

WALLACE TUCKER AND
KAREN TUCKER

William Morrow and Company, Inc.
New York

Copyright © 1988 by Wallace Tucker and Karen Tucker

Grateful acknowledgment is given to W. Forman, C. Jones, and the Smithsonian Astrophysical Observatory; the National Optical Astronomy Observatories; the Mount Palomar Observatory; and the Image Processing Laboratory/Smithsonian Astrophysical Observatory for the use of the photographs in this book.

Library of Congress Cataloging-in-Publication Data

Tucker, Wallace H.
    The dark matter : contemporary science's quest for the mass hidden in our universe / Wallace Tucker and Karen Tucker.
        p. cm.
    Bibliography: p.
    Includes index.
    ISBN 0-688-06112-5
    1. Dark matter (Astronomy)   I. Tucker, Karen.   II. Title.
QB791.3.T83 1988   87-33763
523.1'12—dc19   CIP

Printed in the United States of America

First Edition

1 2 3 4 5 6 7 8 9 10

BOOK DESIGN BY KATHIE PARISE

For the Curious Ones

Prologue

Imagine a mural that hundreds of painters are painting on a limitless wall. Parts of the mural are almost complete, except for a few details—a shadow, another blade of grass, a hint of movement—that are carefully being added by several large groups of painters. These details are important because they give the mural its authenticity and authority. They make it comprehensible and believable and beautiful at any given moment. Except on the boundaries. There, other painters are at work. They extend the mural into the unpainted area by adding paint, stroke by stroke, until gradually a new image is formed. But as the wall has no limit, the mural will never be complete.

The scientific world view is like such a mural. At any one moment it presents a comprehensible picture of the world into which the details are continually being added. And, like the mural, its boundaries are vaguely defined. Science is the process by which details are added and boundaries extended. And like the process that leads to great art, science has its own internal logic that dictates how the parts are added to the whole.

The rules that limit the scientist in carrying out the scientific process—repeatability of observations and experiments, testability of theories—are usually thought of as being far more restrictive than those that guide the artist. They no doubt are. But these rules are not cut and

dried, especially on the boundary. There, individual scientists add a few brushstrokes, some tentatively, some
boldly. Many of these strokes will be repainted almost immediately. Others will stand for a while longer as the scientific community tries to determine whether or not they
fit into the mural correctly.

This book is about science on the boundary. It describes the search for the solution to the mystery of dark
matter in the universe. We have tried to clear a trail
through the scientific literature, leading the reader by the
important points of the dark-matter mystery. Analogies
and examples are used to help clarify some of the tech-
nicalities and a certain restraint was exercised to protect
against too much detail on topics that may hold a fascina-
tion for the specialist but not for the general reader. We
have used interviews and personal experiences to present
the current status of research and opinion on the problem,
and to illustrate the process of science in our time.

The scientific process has a richness and life of its own
that should be emphasized when delving into a puzzle
such as this. Personalities of individual scientists play a
large role in shaping the provisional picture. Strong personalities can have undue influence, sometimes leading
other scientists to fill in the details of the wrong image,
while the correct image is neglected or overlooked.
Clearly, science as well as art is a human endeavor. What
scientists think, how they feel, their preconceptions and
prejudices, their style and their energy, play a role in expanding our knowledge.

Scientists are, after all, people. For better or worse they
do not always think scientifically, even when they are
making discoveries and solving problems. But they do
solve problems and make discoveries. The mystery of the
dark matter in the universe reveals the quintessentially
human trait of curiosity, the restless quest for more

knowledge, and the unceasing desire to push forward into the unknown.

The expansion of the mural has begun. Strong colors have been applied to the walls. What do the mysterious patterns suggest? Are they familiar or strange? Will they change our perception of the mural as a whole? As the mural of our knowledge of the universe expands, we know more. But the boundary between the known and unknown also expands. The mystery of dark matter in the universe is a frontier, but it is by no means the last frontier.

Acknowledgments

We wish to express our admiration and gratitude to all the astronomers and physicists who have worked to bring to light the existence and nature of the mysterious dark matter in the universe. Without their efforts this book could not exist. In particular we would like to thank our friends and colleagues Yoram Avni, John Bahcall, George Blumenthal, Geoffrey Burbidge, William Forman, Margaret Geller, Stephen Gregory, John Huchra, Christine Jones, Mordehai Milgrom, Philip Morrison, Joel Primack, Vera Rubin, Harvey Tananbaum, and Anthony Tyson, who took the time to discuss with us their research and to patiently answer our many questions. Finally we would like to thank Stuart Tucker for his illuminating illustrations, Sandra Dijkstra for her efforts and encouragement, and our editor, Maria Guarnaschelli, for her enthusiasm and valuable editorial assistance.

Contents

THE
DARK
MATTER

1

Dark Matter

A fable:

Once upon a time a small tribe lived in a "habitat" with walls and roof made of an invisible, transparent material. On the outer edge of the habitat was a moat that was home for some very large and very hungry crocodiles, so the tribe could not leave the habitat, or even touch its walls. They could not touch the roof either because it was too high.

For years they remained unaware of the existence of the roof and the walls, since they could not see or touch them. Sunlight passed unimpeded into the habitat, providing light and warmth. They just took for granted that what they could see in their habitat—some furniture, pottery, a rug or two, a garden, a few goats, birds, bees, flowers, crocodiles, and so on—was all there was to it. This belief was reinforced by further observations. On exceptionally clear days they could see what they believed to be other, similar habitats in the distance.

After a while, though, certain members of the tribe, called Curious Ones, began to suggest that there was more to their habitat than met the eye. Heat generated inside the habitat by sunlight or by fires did not escape readily. The trees outside the habitat were sometimes bent and stripped of their leaves by the wind. They concluded that the habitat had walls made of some invisible, transparent material.

Some of the more outspoken and obnoxious of the Curious Ones were fed to the crocodiles. But others continued to gather evidence on what they called The Mystery of the Invisible Walls. They tried to figure out what they could be made of and how they came to be there. Their ideas would not go away.

A true story:

An intelligent species, called humans, lives in a habitat called Earth. Along with eight other planets, Earth orbits a star called the Sun. Over the years, it became apparent to humans that the Sun and its planets, called the solar system, is part of a much larger, saucer-shaped habitat called the Milky Way galaxy.

Humans called astronomers use powerful optical, radio, infrared, and X-ray telescopes to learn many wonderful things about this larger habitat. It contains several hundred billion stars; some of them are very young, having formed only a few thousand years ago. Some are very, very old, more than twice as old as Earth and the Sun, which are four and a half billion years old. Some have bloated up to diameters a thousand times greater than that of the Sun; some have collapsed to diameters tens of thousands of times smaller than the Sun, some of them have even collapsed without limit to form a bottomless pit, a warp in space called a black hole. Besides stars, the Milky Way contains giant clouds of gas and dust, each one with raw material for thousands of new stars.

For years, humans took it for granted that what astronomers had detected with their telescopes—radiation from stars of all sizes, colors, and ages, high-energy particles, gas, and dust—was all there was to the galaxy. This belief was reinforced by further observations. They could see what they believed to be other, similar galaxian habitats far outside their own Milky Way galaxy. Light from these galaxies reached them from many millions of light-years.

In recent times, though, evidence has been accumulating that there is more to our habitat than what at first meets the eye of the telescope. Astronomers now suggest that as much as 90 percent of the matter in the universe must be in some dark form that has so far escaped detection. How is this possible? A careful analysis of detailed observations made with radio, infrared, optical, and X-ray telescopes have revealed some puzzling inconsistencies in our picture of the universe. Spiral galaxies spin faster than they should. Hot gas is trapped around many galaxies. Swarms of galaxies are moving much faster than should be possible.

There is a growing belief among astronomers that galaxies are embedded in massive clouds of dark matter —that what we see in the Milky Way galaxy and other galaxies is only a small part of what is really there. Does the dark matter exist? If so, does it take the form of black holes, brown dwarfs, or some bizarre type of elementary particle? The mystery of the dark matter has become one of the most important issues of the day for anyone who wants to understand what the universe is and how it works. The dark-matter problem is so pervasive that the solution promises to revolutionize astronomy and cosmology.

An analogy can help illustrate the mystery. Suppose a spacecraft is launched with the intention of placing it in an Earth orbit. Due to a computer error, the rockets burn too long. Our calculations based on the known mass of

Earth and the laws of gravity show that this extra burn will cause the spacecraft to overcome the gravitational pull of Earth and shoot into interplanetary space. But suppose that instead the spacecraft goes into orbit! We would be forced to conclude either that Earth has more mass than we thought and hence a stronger gravitational pull, or that the theory we have used to make the calculation is in error.

This is about the situation astronomers find themselves in when they try to understand the motions of stars in the outer regions of galaxies. On a small scale, there is no problem. They understand to high precision the orbits of satellites as well as the orbits of Earth and the other planets around the Sun. But when they analyze the motions of stars and galaxies on a scale of tens of thousands to millions of light-years (a light-year is the distance that light travels in a year, about six trillion miles), it becomes apparent that something is wrong.

In the past few years radio and optical telescopes have measured the rate at which stars and gas clouds in the outer parts of spiral galaxies are orbiting the center of mass of those galaxies. Optical photographs show spiral galaxies to be graceful pinwheels of billions of stars, with the light falling off steadily from the central to the outer regions. Since the light is produced by stars, we naturally expect there to be fewer stars in the outer regions.

A galaxy is held together by gravity. It is not a rigid, rotating body such as a wheel. The strength of the gravitational pull or force of a galaxy is proportional to the amount of matter in it. Thus the force of gravity should be strongest near the center of a galaxy, where the matter appears to be concentrated. As the matter thins out along the spiral arms, the gravity should decrease. It follows then that the speed of rotation of the stars and gas clouds should decline from the inner to the outer regions.

This is *not* what is observed. More powerful instruments have been used to extend the velocity measurements for stars and gas clouds to the outer regions of spiral galaxies. An analysis of these measurements has shown that the stars are not slowing down; they are moving at the same speed as the ones closer in. The gravity of the visible matter is not strong enough to hold these stars in orbit. They should fly off on a tangent, zipping out of the galaxy.

The outer regions of spiral galaxies should have been slung into intergalactic space long ago. Yet there they are, orbiting just as if nothing were wrong—or perhaps more appropriately, as a red flag, telling us that something is wrong with our understanding of galaxies.

Does the problem lie in our understanding of gravity? Is there some additional force that comes into play over these very large scales, a force that is missing from our calculations of the orbit? This possibility is under active consideration by a small group of astronomers. However, most astronomers are reluctant to scrap the Newton-Einstein theory of gravity, which has served them so well. They prefer to assume that it is correct and explore another alternative to explain the rapid rotation of spiral galaxies.

This alternative approach leads to the conclusion that spiral galaxies contain large amounts of matter that has escaped detection. This dark matter would produce the extra gravitational force that is needed to keep the stars and gas on the outer edges of spiral galaxies.

The observations imply that a substantial part of the mass of spiral galaxies is not concentrated toward the center of the galaxy, as the distribution of light would suggest. Rather, it must be in some dark, unseen cloud of matter that pervades the galaxy and extends far beyond it. The outer regions of galaxies, where the matter is faint

and inconspicuous on a photograph, may actually contain
most of the matter. As much as 90 percent of the mass of
galaxies may be in some dark form. In other words, we are
not talking about a dog that is missing a tail. We are talking about a tail that is missing a dog.

Dark matter is not confined to spiral galaxies. A recent
study of data gathered by an X-ray telescope has revealed
that elliptically shaped, that is, round or egg-shaped, galaxies are surrounded by clouds of hot gas. The gravitational pull of the visible matter in these galaxies is not
sufficient to trap this hot gas around the galaxies. It
should have dispersed into intergalactic space billions of
years ago. Implied by these observations is a dark envelope containing ten times more mass than that comprising
the visible galaxy.

On a much larger scale the motions of galaxies in clusters, some of which contain thousands of galaxies, show
the same peculiarity. The observed speeds of the individual galaxies in the clusters are so great that the clusters
should have flown apart billions of years ago. Once again,
the observations require that at least 90 percent of the
matter in the clusters be in some dark form.

In the past few years, astronomers have discovered that
clusters of galaxies are themselves grouped into superclusters. Studies of the supercluster to which the Milky
Way galaxy belongs indicate that our galaxy is moving in
an orbit around the giant Virgo cluster of galaxies fifty
million light-years away. A detailed analysis of this motion suggests—you guessed it—that this so-called Local
Supercluster contains a much larger amount of material in
some dark form that has escaped detection.

On an even larger scale, some astronomers have suggested that the amount of dark matter in the universe
might be sufficient to affect the future expansion of the

universe. There might even be enough of it to halt the expansion of the universe and cause a recollapse.

Not surprisingly, astronomers have in effect issued an all-points bulletin in their search for the dark matter. Instruments sensitive to virtually every wavelength are being pushed to the limit, and theories are being stretched to the breaking point in what has become one of the most intensive "matter hunts" in history. The issues involved in the mystery of the dark matter are, as we shall see, very large ones. They relate to the fundamental nature of matter, the origin of galaxies, the size of the universe, and the future of the universe.

This is not to say that the issues are too profound and too far out to be of concern to the nonscientist. Theories and observations as to the nature of the universe have always touched the lives of everyone. The constant, reliable rhythm of the night sky, the predictable return of Sirius, Orion, and the Pleiades to winter skies and Scorpio to summer skies, and the grand celestial wheeling of the Big Dipper and Cassiopeia in the north must have set the archaic mind to thinking about grand schemes, divine orders, and gods. When Isaac Newton's laws showed that these motions could be understood in terms of a few formulas, all suddenly became light.

Apart from the immense practical value of Newton's calculus and laws of motion, the work of Newton and his contemporaries transformed the outlook of educated people. Comets, long believed to be portents of doom, were shown to be obedient to the law of gravitation. In the words of Bertrand Russell, "The reign of law had established its hold on men's imaginations, making such things as magic and sorcery incredible." It also allowed men such as John Locke, who was strongly influenced by Newton's work, to question the divine right of kings. He wrote that the proper political system is one of "Men living together according to reason."

In the twentieth century we have Albert Einstein's theory on the equivalence of mass and energy and the quantum theory of matter. These have enabled us to understand the atomic nucleus and have propelled us into the space age. At the same time they have brought us to the brink of extinction. Whether we realize it or not, what is known about the nature of the universe does affect the quality of life in our habitat; indeed, it may determine whether or not we will continue to have a habitat.

We can refuse to accept the ideas of the Curious Ones—certainly sometimes their ideas deserve rejection—or we can throw the Curious Ones to the crocodiles—they probably even deserve this sometimes—but we ignore them at the expense of being unaware of our own destiny.

CHAPTER
2

The Galactic Disk: First Evidence for Dark Matter

The investigation into the mystery of dark matter is in many respects like a trial. Evidence is presented to build a case circumstantially or, much better, by using "eyewitness" observational data. Many different, independent strands of evidence are collected. An attempt is made to independently support each strand and its implications. Then the evidence is subjected to cross-examination by fellow scientists. Sometimes their intent is merely to get more information. More often, though, they take the position of adversaries, usually friendly but not always, as they try to chip away at the case.

The evidence begins close to home in the disk of our galaxy. From there it expands gradually to include the entire galaxy, other galaxies, and collections of galaxies until dark matter ultimately emerges as a mystery of universal significance. The evidence and theory at each level is critically important, like pieces of a puzzle, to any possible solution of the mystery.

The case for dark matter in the universe began in our galactic neighborhood in 1932. Jan Oort, a young Dutch astronomer, was trying to understand how the stars in our galaxy moved. A few years earlier he had shown that the Sun and stars were all part of a flattened, rotating disk that we call the Milky Way galaxy.

The stars move around the center of the galaxy much like the horses on a carousel. Just as the horses on a carousel bob up and down a few times as they move around the center of the carousel, the stars bob above and below the disk of the galaxy a few times in the course of each orbit around the center of the galaxy.

The cause for this bobbing motion is gravity. As the stars move above the disk they are pulled back down by the combined gravitational field of all the stars in the disk; they do not stop in the disk, but overshoot and move below the disk, where they are pulled back up to the disk, and so on. The motion is similar to that of the swing of a pendulum. It is continually pulled back to the low point of its path but, in the absence of friction or air resistance, overshoots.

By measuring the distance that the pendulum swings and the time it takes to complete one swing, it is possible to determine the gravitational force acting on the pendulum. This is the method that Galileo used to first determine the strength of Earth's gravity. It is also the method that Oort used to measure the amount of matter in the galaxy.

There is, however, a problem. It takes millions of years for a typical star to complete one swing above and below the disk of the galaxy. Patience is a virtue, but in this case it is clearly a useless one. We cannot hope to follow a star's motion for more than a tiny fraction of a complete swing. In effect, all we have is a snapshot of one instant in

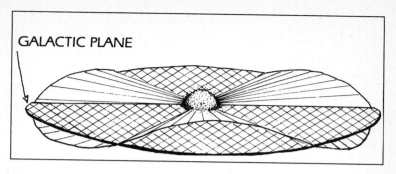

GALACTIC PLANE

As they orbit the center of the galaxy, stars bob up and down like horses on a carousel.

time. We must construct from this snapshot the motions of stars over a time span of many millions of years.

To get an idea of the method that Oort developed for doing this, let us return to the carousel analogy. Suppose you were standing beside a very large carousel, one with hundreds of horses moving so slowly that you couldn't wait around for them to bob up and down even once. But assume that you could make two different observations: a snapshot of the horses to determine the position of each horse at that instant, and a Doppler* radar measurement to measure the speed and direction of motion of each horse.

With this information you could determine several things about the motion of the horses, if you assume that

---

*In 1842, Christian Doppler, an Austrian physicist, showed that the pitch of a moving source of sound waves is different from the pitch of a stationary source. If the source is moving toward you, the pitch increases. If it is moving away, the pitch decreases. A familiar example is the rising then falling of the pitch of a police car's siren as it approaches and recedes. The amount of rise or fall depends on how rapidly the siren is moving toward or away from you. If you know the pitch of the siren when it is stationary, then you can use the observed pitch to work out the speed of the siren.

The Doppler effect applies to radar waves as well as sound waves. Highway patrolmen make effective use of it to measure the speed of vehicles. Their radar gun sends out waves of known pitch or frequency. These waves are reflected back to the patrol car by the moving vehicle. The shift in frequency of the reflected waves is then used to calculate the vehicle's speed.

your snapshot and radar measurements represent a typical instant in the history of the carousel. By studying the positions of the horses, you could deduce that a plausible hypothesis is that the horses are moving in a circular pattern, say counterclockwise.

A prediction of this hypothesis is that the radar measurements should show that the horses on the near side of the carousel are moving from left to right, whereas the horses on the far side are moving from right to left; the horses on the right edge should show a motion away from you, whereas the horses on the left edge should show a motion toward you. By comparing these predictions with the radar observations you could verify that the horses were moving in a circle. By measuring the speeds of horses in various positions you could also determine the speed at which the carousel was revolving.

This is basically the method that Oort used to determine that our galaxy was rotating. The motions of the stars can be determined in essentially the same way as the highway patrol determines your speed. Fortunately stars emit their own light, so there is no need to bounce radar waves off them. If the star is moving toward you, its starlight is shifted to slightly higher frequencies, that is, bluer colors. If the star is moving away, the starlight is shifted to slightly lower frequencies, redder colors. The magnitude of the shift depends on the speed at which the star is moving. With this information plus the data on the positions of stars in the galaxy, Oort was able to discover that the stars are moving in roughly circular orbits around the center of the galaxy. The Sun takes 230 million years to complete an orbit around the galaxy.

Returning to the carousel, we might also notice in our snapshot that the horses are not all in the same positions on their support poles. Some are up rather high on these poles, some are down rather low, and many are in an intermediate position. Furthermore, the horses that are up the highest are moving up or down slowly, if at all. The same would be true of the ones that are lowest down. In contrast, the ones in the middle would be going at fairly high speeds; about half would be moving up, about half down.

All these motions could be explained by a simple theory: The horses are moving up and down on their poles. They move up to the top of their cycle, come to a stop, then start down, pick up speed for a while, then slow down to a stop again as they reach the bottom of their cycle. The details of the observations could be used to verify how far up and down they move and the time taken to complete one cycle.

Oort used an analogous method to study the bobbing up and down of the stars in our region of the galaxy. Once this motion was understood, he could determine the gravitational force that was causing it, and from this force he could determine the amount of matter in our region of the galaxy. He tried to explain this matter on the basis of all different types of visible stars, but was unsuccessful. These stars accounted for only 40 percent of the mass needed to explain the motions of the stars. He expressed little concern with this result, however, and mentioned that when we knew more about very faint stars and interstellar gas, the discrepancy would probably disappear.

Oort repeated these calculations in 1960 using better data. He got essentially the same result. This time he expressed concern. Much more had been learned in the intervening twenty-eight years about the number of very faint stars and the amount of interstellar gas. It no longer seemed possible that this matter could account for the observed distribution and motions of stars above and below the plane of the galaxy. At least a third of the mass of the matter in the neighborhood of the sun was, Oort wrote, "unexplained."

These were strong words for Oort and they spread a sense of unease through the astronomical community. By now, Oort was recognized by many of his colleagues as the greatest living astronomer. Though his proof that the galaxy is a rotating disk was enough to secure his fame, it was only one of many epochal contributions. He was instrumental in launching an entirely new field of astronomy, radio astronomy, which helped to further elucidate the nature of our and other galaxies. He wrote classic papers on the origin of comets, the remnants of exploding stars, the formation of galaxies, the structure of galaxies, and the fate of the universe.

Oort's influence spread far beyond the work in which he himself participated actively. Astronomers were impressed not only with the content of his work, but by his style. They held him up as a paragon for scientists. His style was characterized as a "feeling for the facts," "a quiet rigorous questioning," and, in the words of astronomers Margaret and Geoffrey Burbidge of the University of California at San Diego, as a search for "the truth, the whole truth and nothing but the truth."

So if Oort believed something was amiss, then it probably was. Maybe the universe was substantially different from the way it appeared. Maybe much of the matter in the universe really was hidden in some dark unknown form. But then again maybe it would go away as observations improved.

In 1984 Oort's work was updated in a series of papers by John Bahcall of the Institute of Advanced Study in Princeton, New Jersey. In many respects Bahcall, who is in his early fifties, has had a career similar to Oort's. He has distinguished himself with work on a broad spectrum of astrophysical problems. Also, like Oort, he is a theorist who has worked hard to promote better tools for observers—he has played a major role as one of the most effective advocates of the Hubble Space Telescope.

Bahcall's calculations improved on earlier results in several respects. First, the observational material for the "tracer stars," the ones used to trace out the gravitational force—the horses on the carousel in our analogy—has improved. Second, we know much more about the number of faint stars and the amount of interstellar gas and dust in the galaxy. Complex models describing the distribution of this material have been constructed by Bahcall and others. Finally, modern computers make it possible to determine the gravitational force from these models, and to consider numerous variations of the models. Before it was feasible to make only a few calculations for very simple models.

Bahcall's procedure was basically as follows. He would assume a general model for the galaxy—so many stars of one type, so many of the next, so much interstellar gas and dust, and so forth—that was in agreement with observations. Then he would use this model to predict the detailed distribution of the tracer stars—how many of these stars were in the plane of the galaxy, how many were one hundred light-years above the plane and one hundred light-years below, how many were two hundred light-years above and two hundred light-years below, and so on. He would then compare the result with the observations.

Invariably, no matter how he adjusted his model galaxy, he arrived at the same paradox that Oort had. As long as he used a model galaxy that looked like our galaxy, he could not explain the distribution of the tracer stars. The only way to explain this distribution was to increase the gravity. In order to increase the gravity, more mass had to be added to the galaxy in some dark, undetected form.

Bahcall found that the amount of this dark matter had to be approximately equal to the normal matter—stars, gas, and dust—in the galaxy. Hence, roughly half of the matter in the disk of our galaxy remains unaccounted for. He also found that the dark matter has to be distributed in roughly the same way as the normal matter. It must be included in a disk that has a thickness of about four thousand light-years. The diameter of the disk of our galaxy is about one hundred thousand light-years, so the disk presumably has roughly the same proportions as a trash-can lid.

One potential problem with Bahcall's method is the contamination of tracer stars. If the brightness of these stars is not well known, their distance will be uncertain. This leads to uncertainties in the amount of dark matter. Disagreements between researchers who have used this method to determine the amount of dark matter in the galactic disk usually boil down to the use of different tracer-star populations. Bahcall chose two independent sets of tracer stars very carefully and obtained consistent results for both sets.

Because of the thoroughness of Bahcall's treatment, most astronomers now accept the conclusion that roughly half the matter in the disk of our galaxy is in some dark, so far undetected, form.

The Galactic Disk: Dark-Matter Candidates

What is the nature of the dark matter in our disk? Can it be similar to the matter we observe—gas, dust, rocks, comets, planets, and stars? Will it be possible to observe the dark matter in our galaxy? That will depend on how "dark" it actually is.

Darkness, in astronomy, is measured relative to sunlight. How, then, can a star be considered dark? Stars produce energy by the same process as the Sun—thermonuclear reactions in the interior. A simple rule applies: The more massive a star is, the more efficiently it radiates.

For example, a star that is ten times more massive than the Sun radiates a thousand times more energy, whereas a star that is only a tenth as massive radiates only one thousandth as much energy. This is because the gravity of the more massive stars is greater, so the particles in the interior are squeezed closer together and the nuclear reactions work more vigorously. Put another way, a gram of mate-

rial in a massive star produces on the average more radiation than a gram of material in a low-mass star.

This average efficiency or productivity of matter in
stars may be compared to the average productivity of coal
miners. If a coal mine employs 1,000 workers and produces 16,000 tons of coal per day, the average productivity per worker would be 16 tons. Not all of the employees
work in the mine. Some of them do administrative work
or support work such as maintaining the machines used
in the mining. But the mine would not run without them
so they contribute to the overall productivity of the mine
and are included in our average productivity figure. Likewise, in a star, not all of the particles are involved in the
nuclear reactions that actually produce the energy. However, they play their part by contributing to the overall
gravity of the star that compresses the inner regions and
makes nuclear reactions possible.

Now, consider another coal mine that employs 2,000
workers and produces 128,000 tons per day. The average
productivity is 64 tons per day, or four times higher. Likewise, if a coal mine employing 500 people produced only
2,000 tons of coal per day, then the average productivity
would be 4 tons per day. The average productivity of the
workers would be only one fourth as efficient as the workers in the first example. In this analogy, which has nothing to do with the actual productivity of coal mines, the
average productivity per employee declines rapidly with
the number of employees. It takes more workers to produce a set number of tons of coal per day.

What astronomers are looking for is the celestial equivalent of an inefficient coal mine. They need to find stars or
some type of material in our galaxy that is much less efficient than the Sun at producing electromagnetic radiation.

*   *   *

Electromagnetic radiation is the term for a general class of radiation that includes radio, infrared, optical, ultraviolet, X-ray, and gamma radiation. For our purposes, this radiation can be thought of as waves that are similar to water waves in a bathtub. They have a frequency—how many waves pass a given point per second—and a wavelength—the distance between crests of the wave.

Electromagnetic waves come in all wavelengths, but our eyes are sensitive to only a very small portion of the total range. The wavelength of electromagnetic waves decrease steadily from radio waves, which can have wavelengths of several meters, down through infrared, optical and the others to gamma rays, which have wavelengths of less than a billionth of a centimeter.

Astronomers now have telescopes capable of detecting very small amounts of electromagnetic radiation at every wavelength. Dark matter, then, must be dark not only at visible wavelengths, but at all wavelengths. Astronomers hope that their increasingly sensitive telescopes will eventually detect the dark matter. However, even if the material is so dark that it can never be detected directly, all is not lost. It may still be possible to identify the dark matter indirectly, as we will discuss later in the book.

Is it possible that some form of material observed in our galaxy could account for dark matter?

Interstellar gas can be ruled out. Warm gas with a temperature of a few thousand degrees can be detected by optical telescopes. Hot gas—temperatures of a few hundred thousand degrees or more—can be detected by ultraviolet and X-ray telescopes. Cold gas can be observed by radio telescopes.

Nor can dust be the answer. If the disk of our galaxy was filled with enough dust to account for the dark matter, optical astronomy would not be possible. The galaxy would take on an eerie red glow, like the sunset on days when the atmosphere is filled with smoke, smog, or dust. Only the nearest stars and no other galaxies would be visible.

What about rocks? Comets, asteroids, and rocky planets such as Earth, Mars, Venus, and Mercury? They would certainly be dark enough and we know that this type of material exists. In our solar system the total mass of this debris and the planets is negligible when compared to the mass of the Sun. Even if all the stars in the galaxy had the same amount of planets and debris surrounding them as the Sun, it could not account for more than 1 percent of the dark matter. Also, we know from observations of other stars that they cannot be surrounded by many planets or large amounts of debris. They would show up as either reflected light or infrared radiation.

The only possible place to hide large amounts of rocky material is in the space between the stars. There it could have escaped detection.

However, there are major problems with the rock hypothesis. Astronomers have not been able to come up with a believable theory as to why so much rocky material should exist in the galaxy. Ordinary rocks are composed mostly of carbon, oxygen, and silicon. Since the mass of dark matter in the disk of the galaxy is comparable to the mass of the luminous matter, the rock hypothesis implies that a large fraction of the matter in the galaxy is in the form of these medium-heavy elements. This is completely contrary to the generally accepted theory that 99 percent of the atoms in the universe are hydrogen or helium atoms.

The accepted theory says that all the elements were built up from hydrogen, the simplest element. In the first three minutes of the Big Bang that supposedly initiated the universe as we know it, about 20 percent of the mass of all the hydrogen atoms was converted into helium atoms. Lithium, beryllium, and boron, the third, fourth, and fifth heaviest elements, are thought to have been produced by collisions of cosmic rays with heavier elements in the gas between the stars. All the remaining heavy elements were produced by nuclear reactions inside stars.

The generally accepted reason that heavy elements are so rare—they constitute only about 1 percent of the mass of the visible matter in the universe—is that the universe is still too young. Though it is fifteen or twenty billion years old, there simply has not been enough time for stars to process more than about 1 percent of the atoms into heavier elements. Also, observations that the oldest known stars in our galaxy or any other galaxy have little or no heavy elements strongly supports the idea that the earliest stars condensed from a medium that was almost purely hydrogen and helium.

Can the observations be bent and the theory stretched enough to allow for a galaxy composed mostly of rocks? There is a loophole. The oldest known stars are not *purely* hydrogen and helium. There is a trace of heavier elements. Where did these heavier elements come from? Perhaps these oldest known stars are not the oldest stars after all. There may have been a previous generation of short-lived stars that left behind a trace of heavy elements. Presumably these stars injected some of these into space before exploding.

But is it possible to account for a galaxy full of rocks in this way? Even if the earliest, short-lived stars exploded and ejected most of their mass into space in the form of heavy elements, these elements would then have to form rocks of just the right size to be invisible. This is unlikely. The rock hypothesis would further require that only a small fraction of the heavy elements—a thousandth of a percent or so—was left over to be incorporated into the oldest observed stars. This is very unlikely. Put all these improbabilities together and you have a hypothesis that is extremely unlikely, but not impossible.

Rocks as we know them seem to be improbable candidates for dark matter. Could there be other types of rocks, composed of only the lightest elements, hydrogen and helium? It appears not. Helium can be solidified only at extremely low temperatures and high pressures. High pressures are usually accompanied by high temperatures. This makes it doubtful that helium rocks can exist in nature. If by some unusual circumstance they were produced, they would almost immediately evaporate.

Hydrogen rocks could exist at slightly higher temperatures but they too would evaporate quickly unless they were as massive as the planet Jupiter. But then they would not be rocks, but dark stars. Thus we are led to a search for dark stars.

There are two basic kinds of dark stars: stars that have gone through their evolutionary cycle, used up all their nuclear energy resources, and collapsed to become dim or invisible, and stars that have such low mass that they never have and never will shine brightly. This second category of stars is intrinsically dim, now and forever.

In the first category we have white dwarfs, neutron stars, and black holes. A white dwarf is a star that was once approximately the size of the Sun and has, after the exhaustion of its nuclear fuel, collapsed to form a dense ball about equal in diameter to Earth. They are very dense. A teaspoonful of white-dwarf material would weigh ten tons. But because of their small diameter they are only a fraction of a percent as luminous as the Sun. In recent years astronomers have conducted an intense search for white dwarfs in the neighborhood of the Sun. They have found a few, but only enough to account for 10 percent or less of the dark matter. However, Richard Larson of Yale University has suggested that the dark matter in the galactic disk could be explained by white dwarfs if large numbers of them were produced when the galaxy was very young and then rapidly cooled to invisibility. This would require a revision of our ideas as to how white dwarfs cool.

Neutron stars are formed by the collapse of stars about five or more times more massive than the sun. When the nuclear fuel of such a star is exhausted, its core collapses from a diameter of about a hundred thousand miles to a diameter of about ten miles. The gravitational energy released in this collapse produces a catastrophic explosion called a supernova. The outer layers of the star are blown into interstellar space, leaving behind the collapsed core—a neutron star. The density of a neutron star is roughly ten million times that of a white dwarf. The mass of a teaspoonful of neutron-star matter is roughly equal to that of a pyramid of stone three thousand feet high!

In 1968 this theory of the origin of neutron stars received dramatic confirmation when a neutron star was found in the center of the remnants of a star that had exploded a thousand years before. Young neutron stars rotate rapidly and give off highly periodic pulses of radio emission. They are called pulsars. The pulsar phase appears to last about twenty million years, after which time most neutron stars will become invisible. A very few may become X-ray sources in the rare cases when matter from a nearby companion star falls onto the neutron star.

Studies of the numbers of pulsars indicate that a neutron star is formed about once every thirty years in the galaxy. If this has been happening over the entire lifetime of our galaxy, estimated to be about fifteen billion years, then the galaxy would contain around half a billion neutron stars. The number in the solar neighborhood would be only about a tenth the number of white dwarfs. The denser but smaller neutron stars have about the same average mass as white dwarfs, so they do not appear to be the answer to the dark-matter problem.

The only way out of this argument would be to assume that huge numbers of neutron stars were produced long ago, when the galaxy was forming, or before. The difficulty here is that neutron stars are formed in supernova explosions. These explosions would have injected so much energy into the gas of the protogalaxy that the galaxy might not have formed.

In the unlikely event that the supernova explosions did not blow the protogalaxy apart, the mass ejected in the explosions would pose another set of problems. The theory of supernova explosions indicates that neutron stars are formed from stars with masses between about five and twenty-five times that of the Sun. When such stars explode an amount equal to about one solar mass goes into forming a neutron star. The rest is ejected.

A substantial fraction, greater than 1 percent, of the ejected matter will be in the form of heavy elements. This material will be incorporated into the next generation of stars. A prediction of the neutron-star hypothesis is that these stars—roughly ten billion years old—should have an appreciable heavy-element concentration. Just the opposite is observed. The old stars have only minute traces of heavy elements.

Neutron stars simply have too many problems to be viable candidates for the dark matter in the galaxy.

What about black holes? Black holes are thought to be formed from stars that have masses greater than twenty-five times that of the Sun. When such a star undergoes a supernova explosion, gravitational forces will overwhelm all other forces and the core will collapse. In the words of Robert Oppenheimer and Harlan Snyder, "The star tends to close itself off from any communication with a distant observer; only its gravitational field exists."

The gravitational forces in the vicinity of a black hole are so intense that nothing can escape from it, not light or radio waves or X rays. Nothing. A black hole gives off no light. It makes its presence known only through its gravitational field. Black holes appear to be ideal candidates for dark matter.

The major difficulty with black holes as candidates arises from the size of star needed to form them. The stars that produce black holes have to be extremely massive. The number of such huge stars is observed to be far less than the number of stars that would ultimately become neutron stars. The number of neutron stars is therefore expected to be far larger than the number of black holes, so black holes formed from the collapse of massive stars cannot form as much of the dark matter in the disk as neutron stars. Since neutron stars are themselves unlikely candidates, black holes are even more unlikely.

Collapsed stars, with the possible exception of white dwarfs, appear to be ruled out as candidates for dark matter in the disk of the galaxy. Low-mass stars, in contrast, are still in the running.

Astronomy is a science of superlatives and we are all impressed by grandiosity, so it is not surprising that the astronomical objects that get the most attention from astronomers and the public alike are those that are the "biggest": supergiant stars and galaxies, supernovae, supermassive black holes, and so forth. It is refreshing therefore to see the growing interest in red and brown dwarfs. These lowly objects are so dim that they can be seen only with great difficulty.

Proxima Centauri, the nearest star to us, is an example of a red-dwarf star. Part of a triple-star system, it is a faint companion star to Alpha Centauri A and B, two stars that have been known since prehistoric times and generally considered to be the nearest stars. Proxima Centauri, which was not discovered until 1915, is closer, though, at a distance of just over four light-years.

Proxima Centauri is typical of red-dwarf stars. It has a mass of about one tenth that of the Sun and is much cooler and smaller than the sun. Because of its small mass the gravitational energy of the star is small, so the internal pressure needed to balance gravity is low compared to a star such as the Sun.

The nuclear reactions in the center of these stars do not work as efficiently as those in the Sun, so they are very dim. If the Sun were as dim as Proxima Centauri, it would provide us with the light of only forty-five full moons, or about a ten thousandth of the light we now receive.

Red dwarfs may not be bright, but they are plentiful. Of the ninety or so nearest stars to the Sun that have been classified, two thirds of them are red dwarfs. In other words, there are twice as many of them as all the other types put together! The abundance of red-dwarf stars is part of a trend. The less massive a star, the more plentiful it is. How far does this trend continue? Are stars with very low masses—less than a tenth that of the Sun—the most plentiful stars in the universe? Is this the solution to the dark-matter problem?

Almost two decades ago, Willem Luyten of the University of Minnesota made a survey of the number of low-mass stars in our galaxy. He found that the number of stars starts to decline below about a fifth of the mass of the sun. However, his survey was incomplete for the very faint stars. Other astronomers continued to maintain that many more faint stars would be found when more surveys were made with more sensitive equipment.

In 1982 Ronald Probst of Kitt Peak National Observatory and Robert O'Connell of the University of Virginia used specially equipped telescopes on Kitt Peak in Arizona and in Cerro Tololo, Chile, to search for faint stars. They obtained essentially the same result as Luyten had. In 1983 Gerry Gilmore and N. Reid of the Institute of Astronomy in Cambridge, England, reached the same conclusion in a published report.

Many astronomers remain unconvinced. They feel that the sample of stars observed is still too small. John Bahcall says, "I think that as the observations become better, we will find more very low mass stars."

Anthony Tyson of Bell Laboratories disagrees. He and his colleagues have been conducting very sensitive surveys, using the latest in electronic detector technology. "We have been looking for [low-mass stars] in our data and in star counts. There just aren't any. . . . You can't hide the missing mass in [red-dwarf stars]. They have to be smaller."

Stars with masses less than red dwarfs—less than about a tenth that of the Sun—are called brown dwarfs. They are fundamentally different from the red-dwarf stars. Because of their low masses, the interior temperatures of these stars are so low that the nuclear fusion of hydrogen cannot occur. They glow dimly only because of the heat generated by their slow gravitational contraction. An analogous process is the release of energy by a pile of rubble from a collapsed building as it continues to settle. This energy is released in the form of heat or the sound of a chunk of debris settling to a lower position.

Many astronomers have serious reservations about the existence of large numbers of brown dwarfs. Richard Larson of Yale University, one of the world's leading authorities on star formation, has voiced these reservations.

"I don't think there's any suggestion that unseen matter could be in low-mass stars," he says. He goes on to point out that the observations indicate the trend toward more stars at lower masses does not continue indefinitely. Although there are many red-dwarf stars, no brown dwarfs have been conclusively identified.

In spite of this, some astrophysicists still favor the brown-dwarf hypothesis. They speculate that fifteen billion or so years ago, when our galaxy was forming, the conditions were just right for the formation of trillions of brown dwarfs. These ideas suggest that the trend toward fewer brown dwarfs with decreasing mass will turn around at a certain point. This could mean that many more very low mass brown dwarfs exist than the observations so far suggest.

Brown-dwarf stars are cool and emit most of their energy at infrared wavelengths. These wavelengths are slightly longer than that of visible light. Since most of the infrared radiation is absorbed by the atmosphere, searches for brown dwarfs are best performed in space.

From January through November 1983 the Infrared Astronomical Satellite carried out a search for brown dwarfs. They have found none yet. There is still a lot of data to be analyzed and they could just have been unlucky—the thing you are looking for always seems to be in the last place you look—so they are continuing the search. In the meantime, the evidence bearing on brown dwarfs as candidates for the dark matter continues to be negative.

In summary, none of the commonplace candidates for dark matter in the galactic disk have received any support from the observations. Gas, dust, neutron stars, and black holes have been ruled out. Rocks and comets are extremely implausible. Brown dwarfs and possibly red or white dwarfs are still in the running. They are hard to detect and it is difficult to imagine any other reasonable explanation for the dark matter.

This is not to say that astrophysicists have been unim-

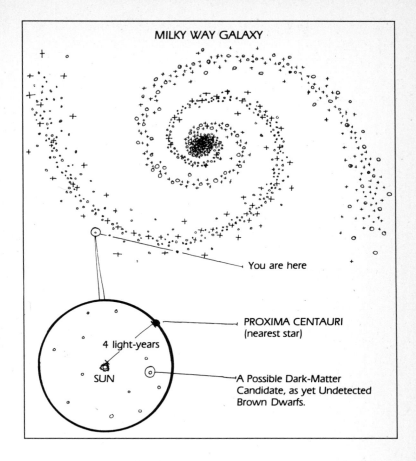

If brown dwarfs are the dark matter in the disk of the galaxy, many of them should be found between us and the nearest star.

aginative. A few have suggested that a revision of the laws of gravity is in order—more on this in a later chapter. Others have extended the search beyond the commonplace. They have followed their theories back to the beginning of time in hopes of finding a form of dark matter that might

explain the observations. There they have found particles
that might solve the mystery—cosmions.

Cosmions are particles that some astrophysicists speculate were produced when the universe was very young,
very dense, and very hot. According to these ideas, long
ago all the matter in the universe was compressed into a
sphere much, much smaller than the period at the end of
this sentence. The temperature of the universe at this time
was about 10^{28} (1 followed by 28 zeros) degrees Celsius.

Theories that have been developed in recent years to
study the behavior of matter under these conditions have
shown that numerous exotic particles are produced. Most
of them would be absorbed without leaving a trace in the
expanding fireball. But some, the cosmions, might survive. These survivors would be particles that interact very
weakly with matter. For this reason, cosmions are also
called cosmic WIMPS, for weakly interacting massive particles. They are loners, so to speak. They don't produce or
absorb light or other forms of electromagnetic radiation.
They would not be detectable with telescopes even
though they might be plentiful. In fact, they could be
plentiful enough to produce the excess gravity we observe. As such, cosmions would seem to make an ideal
candidate for dark matter (see Chapter 5). But they cannot
explain the dark matter in the disk of our galaxy.

When a galaxy first forms, it does not have a disk. It is
a more or less uniform, rotating cloud of gas. The rotation
produces a slight flattening at the poles. Over the course
of time some of the matter in this cloud will radiate away
its energy or lose it through collisions with other gas particles. This low-energy matter will drift down to the equa-
torial plane and form a disk. It is like a pendulum that
gradually comes to rest at the bottom of its arc because of
the energy lost to friction.

Cosmions, by their very nature, interact only weakly with the rest of the matter in the universe. This means that they would not lose energy like the normal matter in the galaxy. They would not have collapsed down to a disk. So the nature of the dark matter in the disk remains a mystery.

Beyond the Luminous Edge: A Galactic Envelope of Dark Matter

How pervasive is dark matter in the galaxy? Is it confined to the galactic disk, or does it extend beyond?

The startling, almost incredible answers to these questions have emerged only in the last few years. Three independent lines of evidence have shown that our galaxy is embedded in a huge envelope of dark matter. What we used to think of as our galaxy—the luminous galaxy—is only the tip of the iceberg. The luminous matter represents at most 50 percent of the actual matter in the galaxy. The remaining matter may be concealed in an ocean of dark stars or exotic particles.

The first line of evidence for an envelope of dark matter around our galaxy comes from radio observations. Radio astronomers have developed techniques for measuring the speeds at which clouds of gas are orbiting the center of our galaxy. From these they can deduce the mass of the galaxy.

The technique involves the use of the Doppler effect. Interstellar clouds of gas have distinctive features in their radio radiation. These can be measured and translated into speeds toward or away from Earth. As discussed earlier, this is the same method used by the highway patrol to measure the speed of automobiles.

In effect, radio astronomers have set up a galactic speed trap to measure the speed of hydrogen-gas clouds in our galaxy. If the cloud is moving toward Earth, the radiation is shifted to shorter wavelengths to a degree that depends on how fast it is moving. If the cloud is moving away from Earth, the radiation is shifted to longer wavelengths.

The key to the application of this method is to find distinctive reference features of the radio radiation from interstellar gas clouds. Fortunately. nature has provided astronomers with a number of such reference features in the radiation from stars and clouds of gas.

Almost all the electromagnetic radiation produced on Earth—in light bulbs, and radio stations, for example—and elsewhere in the universe—in stars, interstellar clouds, and so on—is generated by rapid changes in the motion of electrons. Electrons are light, negatively charged particles that make up the outer parts of atoms.

Sometimes electrons get torn away from the atoms and become free electrons. There are few restrictions on the type of motion that free electrons can perform. They may oscillate very rapidly, which produces high-frequency radiation, or very slowly, which produces low-frequency radiation, or at some intermediate rate, which produces radiation at some intermediate frequency.

Electrons that have been captured and are part of an atom, however, are not free to move in an arbitrary manner. Their motions are strictly regulated by the quantum rules that govern the structure of an atom, which restrict the electrons' motions to specific levels. These can be thought of as stair steps. If you wish to move up or down the stairs, you must move from one step to another one. You cannot move to a position between the steps. You might do more than one step at a time, but you would have to go in whole steps. You couldn't go two and a half steps, for example. Likewise, an electron in an atom can only move from one level to another one.

As electrons jump down from one quantum state to a lower one, they give off packets of electromagnetic radiation called photons. The energy of these photons is related to the magnitude of the quantum leap—the height of the step in our analogy. Streams of photons can be thought of as electromagnetic waves. The energy of the photons corresponds to the wavelength of the wave. A high-energy photon corresponds to a short-wavelength wave.

Because of the quantum restriction that electrons must jump from one distinct level to the next, the photons must have very distinct wavelengths that conform to the arrangement of the levels.

The visible radiation from a hydrogen atom, for example, is not spread smoothly over a rainbow of colors from red to blue wavelengths. Instead it is concentrated in a few extremely narrow wavelength bands. For the most part it is limited to very narrow bands of red, blue-green, blue, violet, and ultraviolet light. These are called spectral lines.

A careful study of the pattern of radiation given off by hydrogen atoms reveals the blueprint for the levels of a hydrogen atom. What scientists have found is that the quantum levels of all hydrogen atoms are the same. A hydrogen atom extracted from the water in the birdbath in your backyard will have the same basic structure as one on Jupiter or in a cloud of interstellar gas. Hydrogen is hydrogen, wherever you find it. And you can find it by looking for a set of photons that match the blueprint.

These are just the type of features astronomers need to set up a galactic speed trap. They can use the Doppler effect on specific features of the radio waves produced by a cloud of hydrogen gas to determine the velocity of the cloud.

In 1944 one of Jan Oort's colleagues at Leiden Observatory, Henk van de Hulst, identified a feature that would turn out to be of immense importance for radio astronomy. He showed that an electron jumping from one to another of the lowest levels would produce a spectral line at a wavelength of twenty-one centimeters. This would place it in the radio band. If the galaxy contained large clouds of hydrogen gas—in 1944 no one knew—then a telescope capable of detecting radio waves should be able to detect the twenty-one-centimeter line.

The importance of van de Hulst's work was recognized immediately by Oort and his collegues at Leiden. Unfortunately, the Nazi occupation of Holland at that time prevented the construction of radio receivers of any kind. Also, because of the occupation, communication with the outside world was cut off, so other scientists could not pursue van de Hulst's ideas.

After World War II, however, the news of his proposal spread quickly. In 1951 Harvard University physicists Edward Purcell and Harold Ewen built a radio receiver sensitive to twenty-one-centimeter radiation and detected the line that van de Hulst had predicted. Six weeks later C. A. Muller and Oort at Leiden had confirmed the Harvard result. A month later an Australian team had also detected it. Within a few years radio telescopes around the world tuned to twenty-one centimeters were mapping out the motion and extent of hydrogen-gas clouds in the galaxy.

Oort, as he had done so many times, led the way. He had already discovered that our galaxy rotated like a giant carousel and that it contained dark matter. But he approached his work carefully and he knew the uncertainties inherent in his work of the 1920s. The observations used to demonstrate the rotation of the galaxy were made with optical telescopes. They were compromised by interstellar dust, which absorbs starlight and makes it impossible to observe the distant reaches of the galaxy. His analysis of the amount of dark matter in the galaxy was made uncertain by an ignorance of how much matter could be hidden in cold clouds of hydrogen gas that were invisible to optical telescopes.

Oort saw immediately that radio observations of the twenty-one-centimeter line offered a solution to the difficulties posed by the effects of interstellar dust. Unlike starlight, the twenty-one-centimeter line is not affected by this dust, so virtually the entire galaxy can be mapped. The galactic radio map shows where clouds of hydrogen gas are, how large they are, and how they are moving. This information allowed an accurate determination of both the amount of hydrogen gas in the galaxy and the rotation of the galaxy.

Oort's original conclusions were confirmed. The galaxy is rotating. However, the rotation of our galaxy differs from that of a carousel in one important respect. A carousel is a rigid structure held together by rigid struts. The horses on the outer part of the carousel make one complete revolution in the same time that it takes the horses on the inner part of the carousel to do so.

Our galaxy, in contrast, is not a rigid structure. It is not held together by rigid struts, but by gravity. It rotates more like a fluid object, such as a hurricane, with the outer

The Milky Way galaxy, if seen edge on, would resemble the spiral galaxy NGC 4565 in the constellation of Coma Berenices.
(National Optical Astronomy Observatories)

parts rotating more slowly than the inner parts. This produces the graceful, swept-back spiral appearance.

The degree to which the stars and clouds in the outer parts of the galaxy lag behind those in the inner parts depends on the distribution of mass in the galaxy. By the mid-1960s a fairly detailed model for the distribution of matter in the galaxy had been developed. It was based on an analysis of observations of the luminous matter in our galaxy and in other spiral galaxies thought to be similar to ours.

According to this model, our galaxy has roughly the shape of a fried egg. The yolk of the galactic egg is a central bulge composed of mostly older stars. The white corresponds to a thin disk of young stars, gas, and dust. Surrounding the galaxy, like gnats around an egg, are a few hundred clusters of older stars. They are called globular clusters, because the stars in the individual clusters have taken the shape of a ball or globe.

There was no provision for dark matter. The distribution of mass in the galaxy was assumed to follow the distribution of light. This implied that the mass, like the light, is strongly concentrated toward the center of the galaxy. Gravitational pull would then be expected to fall off rapidly from the center. Stars and gas clouds in the outer parts of the galaxy would be pulled around at slower speeds than stars and clouds in the inner part.

The standard model predicted that the orbital speeds of stars and gas clouds should decline steadily at large distances from the center of the galaxy. In this respect it was an acceptable scientific theory. It was testable by observations. It did not, however, pass the test.

In the past few years radio astronomers have shown that the standard model is grossly inconsistent with observations. Using the twenty-one-centimeter hydrogen line and an analogous radio-wave line produced by carbon-monoxide molecules, the orbital speeds of gas clouds have been tracked to distances of about fifty thousand light-years from the center of the galaxy. This is beyond the point where there is any appreciable light from the galaxy. Yet the orbital speeds do not decline.*

If the planets behaved in this way, that is, if Earth orbited the Sun with the same velocity as Mercury, it would quickly fly away from the Sun into interstellar space. The centripetal acceleration of Earth would be too great for the gravitational force of the sun to hold it. Similarly, the gravitational force of the galaxy—if we use only the visible matter to calculate this force—is inadequate to hold the distant, rapidly moving galactic gas clouds. Yet there they are.

What is holding these rapidly moving clouds to our galaxy? Evidently it is the additional gravity provided by dark matter. How much dark matter is required? The radio observations imply that at least 50 percent of the mass is in some dark form.

How is the dark matter distributed? Is it spread out in a thin disk like the luminous matter in the galaxy? Radio observations of the distribution of hydrogen gas on the outer edge of the galaxy provide a clue. The gas extends well beyond the optical disk of the galaxy. On its outer edges, the hydrogen disk begins to bloat up and become thicker.

The observed bloating up of this hydrogen gas means

---

*The fact that the stars and gas clouds in the outer part of the galaxy orbit at about the same speed as those in the inner part does not mean that the galaxy rotates like a rigid body. Imagine that the horses on the carousel are brought to life and placed on a racetrack. If all the horses ran at the same speed, the horses on the inside of the track would finish the race before those on the outside of the track, because they run a shorter distance. Likewise, stars or gas clouds on the "inside tracks" in their galactic orbits will complete their orbits first because they have a shorter distance to travel. This will produce the swept-back appearance of spiral arms, even though all the orbital speeds are comparable.

that the dark matter cannot be confined to a thin disk. It must be distributed in a more or less spherical envelope.

The second line of evidence for an extended envelope of dark matter around our galaxy comes from observations at visible wavelengths. This method, which uses observations of the most rapidly moving stars in the galaxy, has been developed by Bruce Carney of the University of North Carolina and David Latham of the Harvard-Smithsonian Center for Astrophysics in Cambridge, Massachusetts.

To understand their method in terms of our model galactic carousel, we have to assume that the carousel is a little more complicated than we first thought. In addition to the horses that partake in the stately bobbing and orbiting motion, there are a few horses available for daredevils to ride. These horses can zip in toward the center of the carousel, make a pass, and come out again at high speeds. They are connected to the center by strong elastic ropes that pull them back toward the center just when it looks as if they might break free from the carousel.

Is there a danger that these high-speed horses would collide with the other horses in the carousel? No, hardly any. This particular carousel is very large and there are great spaces between the horses. Suppose that it is possible for the riders of these horses to get them moving at extremely high speeds. This could be done, for example, by having someone push the horses or by "pumping" them in roughly the same manner that you can pump up a swing.

If the horses are moving too fast, they will break the elastic ropes and fly off the carousel, giving the daredevils perhaps more excitement than they bargained for. Suppose you wanted to know how fast is too fast, or in other words, how strong the elastic rope is. You could determine this by studying photographs and radar measurements of the horses.

You could pick out the daredevil horses, not so much by their positions, but by their high speeds. A careful study of the higher speeds would show that there is a limit. For example, your observations might show that there are 50 horses moving 20 miles per hour, 40 moving 25 miles per hour, 30 moving 30 miles per hour, 20 moving 35 miles per hour, and none moving faster than 35 miles per hour.

The sudden drop in the number of horses above 35 miles per hour would give you a strong clue. The number was declining smoothly up to 35 miles per hour, so there seems to be no problem getting them going that fast. Therefore, you would expect at least a few moving between 35 and 40 miles per hour. But there are none. Evidently the rope breaks at speeds just above 35 miles per hour, so that all horses above that speed have flown off the carousel.

Likewise, in our galaxy there are stars that do not partake in the circular motion around the center of the galaxy, but move at high speeds on elongated elliptical orbits in toward the center of the galaxy and back out again. These stars appear to have been ejected from globular clusters. In a sense, they were pumped up to a high velocity by many near collisions with other stars in the cluster. At a certain point, their velocity became so high that they escaped from the star cluster into the galaxy at large.

Carney and Latham collected a sample of 925 of these "high-velocity stars." In their sample they found no stars moving faster than 500 kilometers per second (roughly a million miles per hour) with respect to the center of the galaxy. Stars moving faster than this must have broken their bonds with the galaxy and escaped into intergalactic space.

The maximum speed, or escape velocity, from the galaxy is related to the gravity of the galaxy. The high escape velocity found by Carney and Latham suggests that the gravity of the galaxy is larger than the amount of luminous matter implies. The extra gravity needed to explain their observations must be supplied by extra matter. Their results point to the galaxy being embedded in an envelope of dark matter.

The use of high-velocity stars to determine the mass of the galaxy is subject to uncertainties. These arise from the fact that we are on the galactic carousel. We are orbiting the center of the galaxy at a speed of roughly half a million miles per hour. This motion must be subtracted from the observed motion of the high-velocity stars to get their actual motion. When all the uncertainties involved in this procedure are taken into account, Carney and Latham estimate that between 50 and 90 percent of the mass of the galaxy is in some dark form. This estimate agrees with that from radio observations.

How extensive is the envelope of dark matter? Measurement of groups of stars that orbit our galaxy at very large distances provides a third line of evidence. These star groups include globular clusters composed of hundreds of thousands of stars and small galaxies with millions of stars.

The technique is basically the same as that used for the radio observations of gas clouds. First, measurements are made of the orbital speed and the size of the orbit of an object. These numbers are then used to compute the mass needed to keep the object in orbit. There is, however, an important difference.

The orbits of the gas clouds in the disk of the galaxy are fairly well known, whereas those of the distant star groups are not. The gas clouds lie in a thin disk and move in circles around the center of the galaxy. The distant star groups lie above and below the disk of the galaxy. They could be moving in circular orbits or they could be moving in elongated elliptical orbits. Since it takes about a billion years to complete an orbit, it is impossible to tell what shape the orbits take.

This ignorance of the exact orbits could lead astronomers seriously astray. For example, less mass is needed to keep a star group in a highly elongated orbit than in a circular orbit. Or the star group could be slowing down as it approaches the outer limit of its orbit. If astronomers assume that the star group is moving at an average orbital speed they could underestimate the mass required to keep it in orbit. The best that can be done under these circumstances is to take an average. Assume that the star groups have on the average an orbit that is midway between a highly elongated orbit and a circular orbit, and an average speed in these orbits.

This averaging process is fraught with danger. The latest and most complete survey of this type involves only sixteen star groups. How do astronomers know that this sample has an average number of elongated and circular orbits? How do they know that their observed speeds represent the average speeds?

Edward Olszewski and Marc Aaronson of the University of Arizona and Ruth Peterson of the Harvard-Smithsonian Center for Astrophysics were appropriately cautious in their report. They concluded that the dark envelope extends to distances of two hundred thousand or more light-years around the galaxy. The actual mass of the galaxy is somewhere between two and thirty times the luminous mass. Their best estimate was that it is about five times as massive. This means that at least 50 percent, and probably as much as 80 percent, of the mass of the galaxy is in some form of dark matter.

*   *   *

In summary, the following picture of our galaxy has emerged from the discoveries of the last decade. In the center there is a bulge of many old stars. This has a diameter of about twenty thousand light-years. Connecting on to this bulge is a thin disk of young stars, gas, and dust, which has a diameter of about one hundred thousand light-years. The material in this disk is largely concentrated into four spiral arms. Our sun is located in this disk. When we look at the Milky Way spread across the night sky, we are looking through the plane of the disk and we are seeing the accumulated light of hundreds of millions of stars.

The dark matter revealed by Bahcall's analysis is concentrated in this disk. Still farther out, the concentration of stars and dust in the disk drops. The disk at this point is mostly hydrogen gas; it becomes much thicker as it extends to a diameter of at least 150,000 light-years. Surrounding this is a roughly spherical envelope composed of a scattering of globular star clusters and dark matter. Most of the mass of the galaxy—perhaps 80 percent—is in this dark envelope.

The stakes have been raised in the dark-matter mystery. Astronomers are not merely looking for a few dim stars that have slipped through their telescopic network. As much as 80 percent of the matter in the galaxy is unaccounted for. Until we can understand what it is, we cannot pretend to understand what our galaxy is really like.

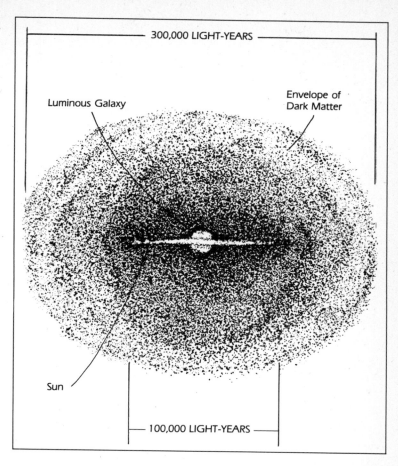

Many lines of evidence indicate that our galaxy is embedded in an extensive envelope of dark matter.

Cosmions: Hot and Cold

Observational evidence and analysis point to a large envelope of dark matter surrounding our galaxy. Now astrophysicists are faced with the task of finding likely physical candidates to explain that evidence. As we have seen, dwarf stars may be able to explain dark matter in the disk of our galaxy. But what about the dark matter in the envelope? Here the conditions are even more extreme, and bizarre dark-matter candidates may be required. As we saw in Chapter 3, this has led astrophysicists into the highly theoretical world of cosmions, where cosmology—the study of the origin, structure, and evolution of the universe—and elementary-particle physics meet.

According to the currently favored Big Bang theory for the origin of the universe, the early universe looked very different from the way it does today. When it was less than a million years old, there were atoms of hydrogen and helium, but little else of the world as we know it. No people, no plants, no elements heavier than helium, no rocks, no earth, no sun, no stars, and no galaxies. Just hydrogen, helium, photons, neutrinos, and perhaps some peculiar form of matter that we now detect as dark matter.

Neutrinos are particles of little or no mass that are known to be produced in some nuclear reactions, just as certainly as it is known that photons are produced in certain nuclear reactions. The conditions believed to exist during the first few minutes of the universe were right for these nuclear reactions to occur, so astrophysicists have predicted that the universe is filled with a sea of primordial photons and neutrinos.

The photon sea has been detected with microwave antennas. Photons produce an exceedingly uniform microwave static that is called the microwave background radiation. The discovery of this radiation by Arno Penzias and Robert Wilson of Bell Laboratories is the strongest and most fundamental evidence to bear out the Big Bang theory. The uniformity of the microwave background indicates that the universe was extremely smooth when it was a few million years old.

The neutrino sea has not been detected and probably never will be. It is not for lack of neutrinos. In the time it takes you to read this sentence, about 100,000,000,000,000,000 primordial neutrinos will have passed through your body, and through the earth. By the time the day is over they will be on their way out of the solar system. If they happen to hit a star they will, in all probability, go right through it too. Primordial neutrinos interact so weakly with the rest of the universe that they might just as well not exist.

Unless the neutrino has mass. Then most of the mass of the universe could be in the form of neutrinos. Not much mass is required—just one ten millionth the mass of the proton, or even a little less—because the Big Bang theory predicts that there are about a billion times more neutrinos than protons in the universe. Neutrinos could be the dark matter around galaxies. They certainly meet one requirement—they are truly dark. In 1980 scientists in the United States and the Soviet Union reported experimental results that suggested neutrinos do indeed have a mass in the right range.

The solution to the mystery of the dark matter in galactic envelopes appeared to have been found not in an observatory looking deep into space, but in a detector underneath a nuclear reactor. The celebration banquet had hardly begun, though, before the ghost of failed theories appeared in the form of two bands of skeptics.

Some physicists formed one band that questioned the interpretation of the experimental results, which have yet to be confirmed. The other band, comprised of astrophysicists, complained that the neutrinos are too hot, that is, they are moving too rapidly. This can cause problems. A universe dominated by fast-moving neutrinos would not look the way ours does. Although neutrinos still have their champions, most astrophysicists began searching for other candidates.

The search led them further back in time, before the first few minutes, before the first few milliseconds, to a time when the universe was less than 1/100,000,000,000,000,000,000,000 of a second old, and into the jungle of theoretical elementary-particle physics, a surrealistic landscape where particles pop in and out of existence like daydreams and the twinkling of an eye is forever.

That this fleeting world of the ultrasmall should have anything to do with cosmology, where distances are measured in mega–light-years and time in eons, is at once strange and obvious. On the one hand, how could the evolution of anything as complex as the universe possibly be influenced by anything so simple as an elementary particle? On the other hand, how could it not be, since elementary particles, subatomic particles regarded to be the irreducible constituents of matter, are the ultimate building blocks of the universe? This idea has forged a coalition of cosmologists and elementary-particle physicists that has invigorated both fields of research. The particle physicists see the early universe as a "one and only time" laboratory, where conditions existed that cannot ever be reproduced in accelerators on earth.

The high temperatures and densities thought to have existed then would have been ideal for testing some of the boldest theories of the nature of elementary particles and the forces that govern them. Was there originally only one type of particle instead of the many that exist today? Are the weak and strong nuclear forces, as well as electromagnetism and gravity all different manifestations of one single superforce? By searching for ethereal particles that were relics of the early universe, elementary-particle physicists hope to answer these questions. And cosmologists hope that the relic particles will help solve the mysteries of the dark matter and the origin of galaxies, and answer the question as to whether the universe is finite or infinite.

George Blumenthal, an astrophysicist, and Joel Primack, a particle physicist, both from the University of California at Santa Cruz, are representative of the new coalition. Primack and physicist Heinz Pagels of the Rockefeller Institute claimed that, according to one class of elementary-particle theories, a cosmion called a gravitino would have been produced in large quantities in the early universe. They would have been invisible and moving more slowly than neutrinos, so they would seem to be a good candidate for dark matter. Primack approached Blumenthal, who was interested, and a collaboration began. Together with Pagels they wrote a paper in which they concluded that a universe full of gravitinos could produce galaxies while avoiding the difficulties associated with a universe full of low-mass, fast-moving neutrinos.

The idea is analogous to designing the enclosures for animals at a zoo. Those that are, like elephants, poor jumpers can be confined by relatively low walls and narrow moats. In contrast, good jumpers, such as gazelles, require high walls and broad moats. The general rule is that the better the jumper, the higher the walls and the broader the moats that are needed for confinement.

A similar rule applies to particles, with gravity playing the role of the confining walls. Faster particles require stronger gravitational fields for confinement than do slower ones. In terms of the formation of galaxies, this translates into the requirement that only galaxies with very large masses, and hence very strong gravitational fields, can form from clouds of rapidly moving particles. Otherwise the particles would escape from the collapsing cloud, and the galaxy would evaporate before it ever had a chance to form.

Blumenthal, Pagels, and Primack found that, for hot dark matter such as neutrinos, the size of the clump that would collapse is ten to a hundred thousand times more massive than that of a large spiral galaxy like our Milky Way galaxy. Even taking into account the dark matter around our spiral galaxy, this number is far too large. The only way it could be relevant to the actual universe was if these huge clumps later fragmented into galaxies.

This leads to a conclusion based on the hypothesis that dark-matter envelopes around our galaxy and other galaxies are composed of neutrinos: The structure in the universe was formed from the top down. At first the universe was a uniform, featureless fireball. As this fireball cooled, huge clouds of neutrinos and ordinary matter started to form rather late, around ten billion years ago. These clouds then fragmented into thousands of smaller clouds. Eventually these would become galaxies, having fragmented into even smaller clouds that produced stars.

Observationally this should show up in the following way: Galaxies would always be in clusters. But this is not observed. The age of distant galaxies and of very old stars in our galaxy indicate that stars and galaxies existed for more than a billion years *before* clusters of galaxies could have formed in a universe full of hot dark matter. Another problem is that very small galaxies, called dwarf galaxies, should not have enough gravity to keep an envelope of hot dark matter, just as the moon does not have enough gravity to keep an atmosphere. Yet dwarf galaxies appear to have envelopes of dark matter (see Chapter 7).

The gravitino seemed to solve the first problem. Since gravitinos would be moving more slowly than neutrinos, galaxy-size clumps of material would form before the larger clumps. However, soon after Blumenthal, Pagels, and Primack published their gravitino paper, the elementary-particle theory that had implied the existence of the gravitino fell from favor and was replaced by another theory, which implied the existence of yet another particle, the photino.

The properties of the hypothetical photinos make them a candidate for dark matter. They would have interacted very weakly with the rest of the matter in the universe, so they would neither produce nor absorb electromagnetic radiation. They could have been produced copiously in the very early universe and have had enough mass to explain the dark matter.

Photinos are cold dark matter, that is, theoretically they move more slowly than gravitinos and much more slowly than neutrinos. This means that clumps of a wide range of sizes could form in a universe full of photinos. A detailed analysis indicates that the first clumps to form would have masses ranging from about one hundred million suns to about a trillion suns. This is very appealing, because it is consistent with the wide range of masses of galaxies observed in the universe. It is also consistent with the observation that galaxies are older than clusters of galaxies.

Photinos are one of the favorite dark-matter candidates. But do they really exist? Are they merely one more product of the fertile imaginations of the elementary-particle physicists that will disappear when yet another elementary-particle theory becomes popular? In a sense, it has already happened.

A particle called the axion had been proposed as a solution to a problem in elementary-particle physics in 1977, long before the photino. No mention was made of using it to solve the dark-matter problem, however, until word of that filtered into the elementary-particle-physics community. Then particle physicists began to think about it and, sure enough, they calculated that the axion would have just those properties necessary to solve the mystery of the dark matter in galactic envelopes. It was weakly interacting, it had mass, and it should have been produced in abundance in the early universe.

Astrophysicists have understandably become cautious about hitching their wagon to any particle-dark-matter horse. Instead, they now use the generic term *cold dark matter* to include any candidate that the particle physicists might come up with that is cold, dark, and material. Then they make their calculations as to the way in which galaxies and other structures form in a universe filled with cold dark matter and they compare this hypothetical universe with the real universe.

This was the approach taken by Blumenthal and Primack, together with their U.C. Santa Cruz colleague Sandra Faber and Martin Rees of Cambridge University, when they wrote an influential review paper, "The formation of galaxies and large-scale structure with cold dark matter." In it they concluded, "We have shown that a Universe with about ten times as much cold dark matter as baryonic [that is, normal matter made of protons, neutrons, and electrons] matter provides a remarkably good fit to the observed Universe."

That was in October 1984. By the summer of 1985 doubts about the cold-dark-matter hypothesis had begun to surface. The problem, ironically, is that cold dark matter makes galaxies too efficiently. They should show up everywhere. Yet they do not.

Surveys of the distribution of galaxies in the universe have shown that the universe, like Swiss cheese, is full of holes. The holes or voids are tens of millions of light-years across, and contain, as far as anyone can tell, nothing. Yet, according to the simplest form of the cold-dark-matter hypothesis, the voids should contain at least a few hundred galaxies. The advocates of cold dark matter are faced with the age-old problem in science: Do they give up their ideal that a correct theory must be simple and beautiful or do they complicate their theory?

For the time being, they are changing the theory so that it works. Cold dark matter, despite its problems, remains the most popular candidate for dark matter in the envelopes around galaxies. But because of the problems, the neutrino partisans have not given up hope. The ultimate solution may not be simple and beautiful. It may involve more than one type of dark matter. Before we can assess this somewhat "ugly" hypothesis and show why it may be necessary, we must examine much more evidence. This evidence will reveal the universality and complexity of the dark-matter mystery.

The spiral galaxy NGC 5364 in the constellation of Canes Venatici
(National Optical Astronomy Observatories)

CHAPTER

The Turning Point:
Dark Matter in
Spiral Galaxies

Successful prosecutors know that a case is made much stronger if a pattern of misdeeds can be established. It is easier to cast doubt over evidence related to one isolated incident than several incidents. It is also possible to overlook an isolated incident as the product of extenuating circumstances, but not so a pattern of criminal activity.

In the same way, a scientific case becomes much stronger if a pattern of phenomena can be established. An isolated instance of a strange phenomenon is intriguing, but it is difficult to know how much importance to attach to it. Usually a few scientists will study it, while the rest merely take note of it and go on about business as usual. If, however, evidence can be developed for a pattern of strange phenomena, scientists will sit up and take notice. A revision of basic ideas may be in order.

The work of Vera Rubin on the rotation of spiral galax-

ies represents such a turning point in the history of the
dark-matter mystery. She and her colleagues have shown
that the envelope of dark matter around our galaxy is not
unique. It is not even rare. Spiral galaxies, no matter what
size, are embedded in dark matter.

Rubin's work is so important that it is worthwhile to
go into some detail as to who she is, what she did, and
why. The why gives insight into the nature of scientific
discovery. Many scientists are constantly following the
hot topics. They go where the action is. Vera Rubin was
not one of them. Because of her personality and her special circumstances, she consciously avoided the action.
Yet she made one of the greatest discoveries of her time.
She chose to study what most astronomers considered to
be a rather dull problem—the rotation of spiral galaxies.

It is a straightforward, if painstaking, matter to measure the rotation of a spiral galaxy. It can be done with
radio telescopes by observing the Doppler shifts of the
twenty-one centimeter line from hydrogen gas, or a similar line from carbon-monoxide molecules. Another
method uses optical telescopes to measure the Doppler
shifts of several spectral lines that stars and gas produce at
optical wavelengths. This is accomplished by means of a
spectrograph.

A spectrograph is a device that spreads the light enter-
ing the telescope into its spectrum of wavelengths, a broad
rainbow of colors running from violet at the short-wavelength end to red at the long-wavelength end. Superim-
posed on the rainbow will be the spectral lines that are
produced by the movement of electrons in the atoms of
particular elements. The position of the lines in the spectrum indicates what their wavelength is. This corresponds
to the change in energy from one quantum level to another
in a particular atom, analogous to the change in height
from one stair step to another.

If an electron absorbs a photon it jumps up a level and takes light out of the spectrum, producing a dark absorption line. If an electron jumps down a level, it emits a photon and adds light to the spectrum, producing a bright emission line. Whether a spectrum has bright or dark lines depends on the conditions in the source. Stars typically exhibit absorption lines whereas hot clouds of gas exhibit emission lines. In either case the positions of the lines are unique—only one type of atom can have a line of a particular wavelength—so the element responsible for a given line can be identified.

A spectrograph makes a photograph of a given galaxy's spectrum, together with a photograph of a reference spectrum from well-measured laboratory sources. By comparing the two photographs, the Doppler shifts of the spectral lines can be computed. This information can then be used to analyze the rotation of the galaxy. Once the rotational velocity of a galaxy at a certain radius from the center of the galaxy is known, the mass of the galaxy inside that radius can be calculated.

The method is fairly simple, but putting it into practice is not. The difficulty is that the nearest galaxies are millions of light-years away. The overall amount of light captured from these galaxies is sufficiently great that a nice image can be made with a relatively short exposure. However, to measure the Doppler shifts and hence the rotation of a galaxy, we must accurately measure the light that falls into the spectral lines, each of which can contain a very small part of the overall light. For example, the first convincing measurement of the rotation of a galaxy—the Andromeda galaxy, the nearest large galaxy to the Milky Way—required two eighty-hour exposures on the sixty-inch telescope on Mount Wilson!

This was in 1917. Though the one-hundred-inch telescope was soon put into use on Mount Wilson and detec-

The rotation of the outer parts of the large spiral galaxy in Andromeda indicates that it is embedded in an envelope of dark matter. (Palomar Observatory photograph)

tors improved somewhat, it wasn't enough to encourage astronomers to rush to the mountaintops to measure the rotation of more galaxies. Astronomers are uncommonly patient and tenacious people, but even they found the prospect of spending over one hundred hours on a telescope to get the rotation of one nearby galaxy daunting.

What few measurements were made tended to be of the bright central regions of galaxies, where the exposure time was less. As late as 1959, the rotation of only eight galaxies had been studied in any detail. Gerard de Vaucouleurs of the University of Texas expressed the prevailing opinion among astronomers when he wrote in a review that, outside the inner region of the galaxy, "the rotational velocity decreases with increasing distance to the center and tends asymptotically toward Kepler's third law." In other words, he concluded that the galaxies rotated in the same way that planets orbit in the solar system. This is just what astronomers had expected to find; since the light from these systems is concentrated in the central regions, most astronomers expected to see the rotational velocities fall off with distance from the center, so that is how they interpreted their data.

Oort, characteristically, was not misled by false expectations. Almost two decades before, in 1940, he had analyzed data gathered by two other astronomers on the distribution of light and the rotational velocity of the galaxy NGC 3115. In an article published in *The Astrophysical Journal,* he concluded that "the distribution of mass in the system appears to bear almost no relation to that of light." He went on to warn, "The total mass of the system cannot yet be estimated; it should be at least 50 billion solar masses but may be much larger."

A few years later, in 1946, Oort broached the idea that some galaxies might be embedded in massive dark halos of red- or brown-dwarf stars. In an article in the *Monthly*

Notices of the Royal Astronomical Society he stated that the outer regions of NGC 3115 must be "made up of extremely faint dwarfs," which should be numerous.

NGC 3115 is not a spiral galaxy with well-defined spiral arms that are populated with gas, dust, and bright young stars. Rather, it has the form of a featureless disk of old stars. The absence of gas, dust, and young stars gives it many similarities to *elliptically* shaped galaxies that have these latter properties.

In 1954 Martin Schwarzchild used newer observations to repeat Oort's calculations. He confirmed Oort's conclusion that the mass of the galaxy was much larger than one would estimate in assuming that it was composed only of normal-type stars. He found a similar result for an elliptical galaxy, but not for two spiral galaxies that he studied.

He concluded that elliptical and disk galaxies are for some reason fundamentally different from spiral galaxies in that they have a large population of extremely faint dwarf stars. This conclusion was accepted by de Vaucouleurs, who repeated it in his review, and by virtually every other astronomer.

The reason most astronomers accepted uncritically the work on the rotation and mass of spiral galaxies was probably that it fit in with their preconceptions. More data were obviously needed, but who wanted to spend long hours at the telescope proving the obvious? Practically no one. Until Vera Rubin came along.

Vera Rubin turned the astronomical community on its ear by showing that spiral galaxies do indeed have massive dark halos. She made this discovery because, among the possible avenues of research, she "took the one less traveled."

Vera Rubin didn't know that her career would turn out

as famously as it has, but she had known for a long time that she would have a career in astronomy. She is one of those rare people who has known since childhood what she wanted to do when she grew up.

"I became interested in astronomy when I was twelve," she relates, "because my bed was right under a window. . . . At night I watched the stars cross the sky. It was fascinating, the motion of the stars through the night, and the change with the seasons. I got so interested that I found it hard to sleep. . . . To find out what was going on I checked some books out of the library and built a telescope."

Although a professor later told her that "people who get interested in astronomy at a young age and build telescopes are the worst prospects," she majored in astronomy at Vassar College. She chose Vassar because it was one of the few colleges that offered an astronomy degree. After Vassar, she received a master's degree in astronomy at Cornell. There she met and married Robert Rubin, a graduate student in physics. Before attending Cornell she had applied to Princeton University.

"I wrote them a letter asking how I could qualify [for the doctoral program]. I got back a letter from the dean saying that 'we do not accept women, therefore you cannot qualify'!"

The difficulty of being a woman in science is a continuing concern for Vera Rubin. She has been actively involved in trying to make the situation better. She serves or has served on the Council for American Women in Science, an American Physical Society Panel on Faculty Positions for Women Physicists and Astronomers, and the Council on Women of the American Astronomical Society. But she is not optimistic.

"I don't see it getting any better," she said. "That's really what bothers me. I have only one daughter [Judith

Young]. She is an astronomer, a good one, and she's had a hard time. . . . The first conference she went to she told me she was the only woman there. I could only think, this is one full generation later and things haven't gotten any better. . . . She had an honor's degree from Harvard, yet she had a hard time getting into graduate school. She eventually got a Ph.D. in cosmic-ray physics, and she does beautiful work [good enough to earn her the Annie J. Cannon Award of the American Astronomical Society for a woman who makes distinguished contributions to astronomy] but she had a hard time getting a job."

To illustrate just how difficult it can be for a woman to get a job in science, she related to us a recent personal experience at an exclusive dinner at the Air and Space Museum. In the course of the evening, one of her colleagues, the head of a famous astrophysics laboratory, learned that Judith Young was Vera Rubin's daughter. "He said, 'I didn't know she was your daughter. Gee, when she was a graduate student she came and gave a talk. It was one of the best talks we ever heard and we thought she was just great. We wanted to offer her a job but we knew she had a husband and she probably wouldn't . . .' That's the end of the story.

"I left thinking . . . a male graduate student gives a great talk and the people tell him it's a great talk. They offer him a job. Whether he takes it or not he's gotten that much encouragement. She goes—I haven't even told her this story—and gives a talk that they say is one of the best talks that they had ever heard, and that she is such an interesting person. And she never heard a word from them. It's that kind of thinking that's—it's just different being a woman."

In Vera Rubin's case, the difference meant that she would have an unconventional career. After receiving her master's degree she moved to the Washington, D.C., area, where her husband had just accepted a job. She applied and was accepted to the graduate program at Georgetown University, the only place in Washington that offered an astronomy degree at the time.

They offered the degree but they did not have anyone to direct her thesis work, so her adviser became George Gamow of George Washington University.

Gamow was an eminent, though unconventional, nuclear physicist with strong interests in molecular biology and astronomy. A few years earlier he had been a co-author of what would become a classic paper on nuclear reactions in a hot, expanding universe. Also a talented writer of general-interest science books, he popularized the theory that the universe began with an explosion, or "Big Bang."

Gamow was interested in astronomy but he wasn't part of the astronomical community, nor was he in very close contact with it. This was not necessarily a drawback, Rubin feels. It allowed her to keep her individuality, to be free from conventional and strictly fashionable approaches to problems.

"I think the students we are turning out today are incredible. Just phenomenal. But," she continues, speaking slowly and emphatically, "they all learn the same things. They know what is practical and what is not practical and they all ask the same questions. So, it's much harder [to maintain an open mind and an individual style of research], I would think. It's beaten out of you in graduate school. . . . I didn't know enough. I never went through that conditioning."

So Rubin received her Ph.D. from Georgetown in 1954 with her curiosity intact. She stayed there for ten more years, working as a research associate and then as an assistant professor. It was a time when everyone was studying the centers of galaxies. "It was a very popular field," she recalls. "And that made me ask the question, 'What's the anticenter [that is, the region of the sky directly opposite the galactic center] like?'" This was the beginning of her work on the rotation of the outer parts of galaxies, although she admits that "I didn't know enough to say that I was interested in rotation. I was just curious about things."

After eight more or less isolated years at Georgetown, she left for the University of California at San Diego. There she worked with the husband-wife team of Geoffrey and Margaret Burbidge. The Burbidges were two of the most productive astronomers in the world, and were very much in touch with the bustling, vigorous activity in the central marketplace of astronomy.

"That was a very powerful time for me," Rubin declares. "I learned that people were interested in what I had to say. That gave me confidence." The Burbidges encouraged her to continue her work on the rotation of galaxies. They also suggested that she get into quasar research.

Quasars, which are now thought to be the energetic nuclei of galaxies fueled by massive black holes, had just been discovered. Their nature was a total mystery. They were unquestionably the hottest topic in astronomy.

At the time, Vera Rubin was a young astronomer seeking to make her mark, so what did she do?

"I took spectra of quasars, which was what you did in that era."

But she quickly found that the frenetic pace of quasar research was not for her.

"After about one year of this I just couldn't stand it," she recalls. "I only went observing about twice a year and I would get a few spectra. . . . And Maarten Schmidt [a Caltech astronomer noted for his work on quasars] or someone else would call and say, 'Do you have spectra of this?' and I would say, 'Yeah, maybe I have one spectrum.' And he would ask, 'Do you have a red shift?'" (Because of the expansion of the universe, all distant galaxies are moving away from us. This motion shifts the spectrum of the distant galaxies to longer, or redder, wavelengths. The nature of the expansion is such that, the more distant the galaxy, the larger the red shift. Quasars have very large red shifts, which is taken by most, but not all, astronomers to mean that they are very distant.)

Such questions made Vera Rubin uncomfortable. "I would either be forced to say 'Yes' and tell them something I wasn't sure of," she explains, "or say 'No,' in which case they would go observe it, so what I had done was lost. This was not the way she wanted to do research in astronomy. To her, astronomy had always been fun.

"I just didn't find that fun," she says. "I really didn't like telling people results that I wasn't sure of. I couldn't sleep. I decided that I was just going to go off and do a problem that nobody would care about while I was doing it. And hopefully when I was finished, I would show them what I had done and everybody would think that it was fine. They didn't have to think [the work] was great, but it should be fine."

She began to concentrate deeply on the rotation of spiral galaxies. In the space of two years, Rubin published ten papers with the Burbidges, mostly on this subject. Then she returned to the Washington, D.C., area, where she took a position as an astronomer with the Carnegie Institution of Washington. It was a job she had had her eye on for some time.

"When I was working for George Gamow, we had a common meeting spot since he wasn't at Georgetown. . . . He would tell me to meet him in the library at Carnegie. It's a gorgeous place, on a parklike campus. It's lovely. . . . I really sort of decided very early on in my life that that would be a nice place to work. So, I essentially waited until I thought they would hire me and I walked in and asked for a job. They were doing some radio astronomy— Bernie Burke [a radio astronomer who is now at the Massachusetts Institute of Technology] was there, and Kent Ford, an optical astronomer, was there developing an image tube."

Ford's image tube was to play a crucial role in the work by Rubin and her colleagues on the rotation of spiral galaxies and the mystery of dark matter. An image tube is an electronic device that converts a stream of photons captured by a telescope into a more intense stream of electrons that are then focused onto a screen like a TV screen or onto a photographic emulsion. Ford's image tube, which became known as the "Carnegie image tube," was a major step forward in the technology needed for studying the spectral lines of faint galaxies. An exposure that would previously have taken sixty hours could now be done in three.

Over the next ten years, Rubin, Ford, and their colleagues Norbert Thonnard of the Carnegie Institution, David Burstein of Arizona State University, and Bradley Whitmore of the Space Telescope Institute observed a sample of sixty galaxies. They chose a wide range of spiral galaxies: tightly wound spirals with bright central bulges; spirals with loosely wound, bright arms; large galaxies twenty-five times more luminous than the Milky Way, and small ones only a quarter as luminous.

They found, in every case, that as they went to the visible edge of a galaxy the rotational velocity did not decrease with the light but remained constant. It even increased in some cases. The galaxies appear to be spinning faster than is possible, if the visible galaxy gives any indication of its mass. Therefore, they stated: "The con-

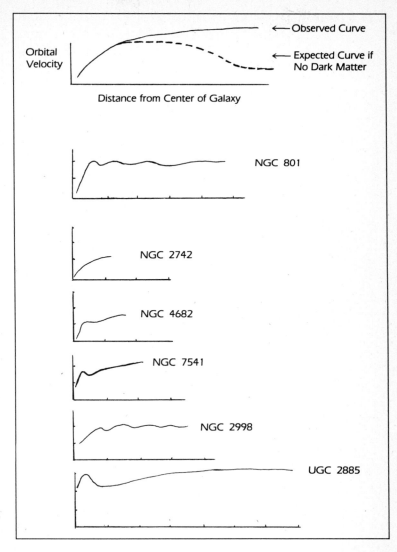

Observations of orbital velocity of stars and gas on the outer edges of galaxies provide convincing evidence for envelopes of dark matter around these galaxies.

clusion is inescapable that nonluminous matter exists beyond the optical galaxy."

Their work showed that our galaxy is not unusual among spirals, with its massive envelope of dark matter. It is typical.

Radio astronomers have also obtained observations of the rotation of many spiral galaxies. These observations, which go well beyond the visible edges of the galaxies, confirm and extend the optical observations. The radio and optical observations together provide convincing evidence that dark matter accounts for 50 to 80 percent of the total mass of the spiral galaxies.

By the early 1980s it was no longer possible to doubt that dark matter exists in spiral galaxies. More than anyone else, Vera Rubin had transformed dark matter from a subject fit primarily for the speculators to a highly visible problem, a hot topic of the type she had tried to avoid twenty years earlier.

CHAPTER

7

Dwarf Galaxies

The evidence for dark matter in and around spiral galaxies is extensive and compelling. A pattern definitely exists. But what about dwarf galaxies? They are much smaller than spiral galaxies. Do they also have envelopes of dark matter? If so, the implications could be far-reaching, because dwarf galaxies are probably the most common type of galaxy in the universe.

The present intense interest in dwarf galaxies is an object lesson in one of the principle articles of faith of many scientists: the inexhaustibility of nature. If you study anything in detail, no matter how prosaic it may appear, you will frequently uncover a gem of breathtaking beauty and value. Think of Newton and the apple, Mendel working with pea plants, Darwin and his finches, Becquerel with his chunks of pitchblende, and Vera Rubin and the spiral galaxies.

Poets have long understood this principle. Wordsworth wrote:

> To me the meanest flower that blows can give
> Thoughts that do often lie too deep for tears.

William Blake wrote:

> To see a World in a Grain of Sand,
> And a Heaven in a Wild Flower.

John Donne was less romantic, but in a way more to the point:

> I neglect God and his Angels, for the noise of a fly, for
> the rattling of a coach, for the whining of a door.

Dwarf galaxies are so inconspicuous that they were not discovered until the late 1930s. They range in brightness (or dimness) from one hundredth of 1 percent to 1 percent of that of a large spiral galaxy such as Andromeda or our own galaxy. In the early 1930s, the conventional view, as expressed by Edwin Hubble, was that all galaxies had roughly the same number of stars and the same brightness. Much smaller galaxies simply did not exist.

Fritz Zwicky, Hubble's Caltech colleague, disagreed. Zwicky, who had a wide-ranging intellect, was involved in a long-running feud with Hubble, the preeminent astronomer of his time.* Zwicky gave two arguments as to why dwarf galaxies should exist.

The first was the principle of the inexhaustibility of nature. Nature, Zwicky argued, is unlikely to allow for

---

* In addition to his work on dwarf galaxies, Zwicky produced seminal work on dark matter, neutron stars, supernova explosions, distant galaxies, and clusters of galaxies. He published important papers on crystal structure, his original field of research, electrolytes, superconductivity, cosmic rays, and civil defense. He obtained fifty patents in the area of rocket and propulsive power, and in 1948 he was awarded the Freedom Medal by President Truman for his wartime work on propulsive power.

At age seventy-four Zwicky was awarded the gold medal of the Royal Astronomical Society of the United Kingdom for his distinguished contributions to astronomy and cosmology. He died two years later, in 1974.

only one type of galaxy. They must come in all shapes and sizes, including dwarfs.

Zwicky's second argument was that if Hubble said dwarf galaxies did not exist, then they almost certainly must exist. This was what he called "the method of negation and subsequent construction." The first step of this method is to look for statements, theories, or systems of thought that pretend to absolute truth, and to deny them. You are almost certain to be correct in doing this, Zwicky maintained, because it is extremely unlikely that anyone knows the absolute truth about anything.

Zwicky, however, was not content with a purely negative approach. His method, true to his principles, was one of negation *and* subsequent construction. This, he emphasized, "has no relation to the general desire of frustrated minds to negate all positive aspects and assets of life." Negation is fruitless unless it opens up new vistas and new approaches to observation and theory for creative investigators to explore.

Having negated the "truth" that dwarf galaxies do not exist, Zwicky proposed to search for them. But Hubble and other authorities at Caltech refused to make available valuable time on the large telescopes to look for something that they had declared couldn't exist.

Undaunted, Zwicky acquired funds to build his own telescope, a small, eighteen-inch reflecting telescope. It was especially designed to search large areas of the sky. Although the telescope had only one thirtieth of the light-gathering power of the one-hundred-inch telescope, Zwicky believed that the smaller telescope's ability to cover wide swaths of the sky would uncover many new phenomena, such as dwarf galaxies, that had been missed with the deep but incomplete coverage of the larger telescope.

The telescope was installed on Mount Palomar in 1936. Within months he discovered two dwarf galaxies, one in the constellation of Leo and one in Sextans. In the course of a four-year survey he found several more.

He also discovered eighteen supernova explosions in distant galaxies that would have otherwise gone unnoticed, numerous previously undetected white dwarfs, and made pioneering investigations of clusters of galaxies. These showed clusters of galaxies to be far more numerous and much larger than previously thought. They also provided some of the first evidence for massive envelopes of dark matter around galaxies. Not a bad record for an eighteen-inch telescope, and it was vindication of Zwicky's belief in the inexhaustibility of nature.

After Zwicky's discovery of dwarf galaxies, Walter Baade of Caltech used the one-hundred-inch telescope to study the dwarf galaxies in Leo and Sextans. He confirmed their indentification and established their distance, which is less than a million light-years from our galaxy.

Since their discovery, dwarf galaxies have not attracted the attention of more than a handful of astronomers. The average distance between stars in a typical dwarf galaxy is about eighty light-years, compared to about eight light-years in the disk of our galaxy. Even the central regions, a crowded, chaotic trouble spot in almost every other type of galaxy, are free of congestion in dwarf galaxies. No explosions, no gas rushing out from the nucleus of galaxy, no evidence for a supermassive black hole. Everything is going along pretty much the way people used to think that galaxies should act, before the discovery of quasars, black holes, and galactic explosions. Dwarf galaxies appear to be very well behaved.

But ironically, because of the dark-matter mystery, dwarf galaxies have become important recently for the same reason that they were previously largely ignored—namely, their

The dwarf galaxy in the constellation of Sextans
(Palomar Observatory photograph)

dwarfness. Since they have few stars, they have a weak gravitational field. This raises interesting questions: Why do
dwarf galaxies exist around our galaxy? Does their existence
imply the presence of dark matter?

Our galaxy is surrounded, appropriately, by seven
dwarf galaxies. Two of these, one in the constellation of
Draco and one in Ursa Minor (the Little Dipper) are only
two hundred thousand light-years from Earth. This means
that they are moving in the outer halo of our galaxy, and
are subject to huge tidal forces.

Tidal forces are familiar to anyone who has visited the
seashore. The ocean tides are caused by the gravitational
force of the Moon on Earth. The side of Earth facing the
Moon is about 3 percent closer to the Moon than the side
facing away, so the gravitational force exerted by the
Moon on the near side is slightly stronger. This imbalance
of forces distorts Earth's shape by a few feet. And this distortion shows up as the high and low tides of the ocean.

The tidal forces of the Moon on Earth are very weak
because the gravitational field of the Moon is weak. The
tidal distortion of the Moon by the Earth is much greater
because the mass of Earth is eighty times greater. Still, it is
fairly weak and distorts the shape of the Moon only
slightly.

However, tidal forces need not always be small. In
some double-star systems, they are enormous. Here the
stars are almost touching; the material in the outer layers
of these stars is torn between the gravitational force holding the material to one star and the gravitational force of
the companion star. If the gravitational field of the companion star is strong enough, it can literally tear the outer
layers off the other star. This is observed to happen in a
number of star systems. The best-known example is probably Cygnus X-1, where the tidal forces, produced by a

star that has collapsed into a black hole, are pulling apart the other star.

In a similar manner, the tidal forces of our galaxy can pull apart a nearby small-dwarf galaxy. Whether this will happen depends on whether or not the dwarf galaxy's gravitational field is strong enough to withstand the tidal forces exerted by our Milky Way galaxy.

Marc Aaronson and, independently, Sandra Faber and Douglas Lin of the University of California at Santa Cruz, showed that the Draco and Ursa Minor dwarf galaxies should have been destroyed by tidal forces. Yet they haven't been. Their gravitational field must be stronger than one would have thought from the number of stars they contain. The solution: They are embedded in an envelope of dark matter. The dark mass required in these dwarf galaxies is ten times the mass in stars. Luminous matter would represent only 10 percent of the total mass of these galaxies.

This analysis of the data has significant consequences for the dark-matter mystery. Aaronson, Faber, and Lin pointed out that the dark matter could not be composed of one type of cosmion, the neutrino.

Neutrinos, though they interact weakly with the rest of the universe, are constrained by the laws of quantum mechanics. One of these laws, named the "Pauli exclusion principle" after its discoverer, Wolfgang Pauli, says that no two neutrinos can occupy the same quantum level. A familiar analog of this principle is a parking lot: No two cars can occupy the same parking space. The extra cars must circle around waiting for a parking space to become empty. Likewise, if all the low-lying quantum levels are occupied, the extra neutrinos are forced to higher energy levels.

If the neutrino has mass, that mass would be restricted

by the Big Bang* model of the universe to be thirty million times less than the mass of the proton. In order to account for the dark matter in a dwarf galaxy, very many neutrinos with this mass must be packed into a small volume. Too many. There would be more neutrinos than available low-energy quantum levels. The extra neutrinos would be forced to higher energy levels. But at the higher energy levels they cannot be confined to the dwarf galaxy. They would escape into intergalactic space. It is not possible to pack enough neutrinos into a dwarf galaxy to account for the dark matter.

The existence of dark matter in dwarf galaxies is one of the most telling arguments against neutrinos as the solution to the mystery of the dark matter. The importance of this result demands that the tidal-force argument be confirmed by some independent method. Though the evidence seems convincing, it is not conclusive. The uncertainty about the mass of our galaxy and the difficulty of applying the tidal-force formula to dwarf galaxies leave open the possibility that the amount of dark matter could be two or three times less. The thrust of Aaronson's latest work has been to close off these loopholes by determining the mass of dwarf galaxies the old-fashioned way, by measuring the velocities of the stars in the galaxies.

This is excruciatingly difficult when applied to dwarf galaxies. Their stars are expected to move slowly because of the weaker gravitational field—any stars moving at moderate speeds would have escaped from the galaxy. Low speeds mean small Doppler shifts, which means they are difficult to measure. This difficulty is aggravated by

---

* As discussed in Chapter 5, the Big Bang theory implies the existence of about a billion times more neutrinos than protons in the universe. If each neutrino has a mass more than 1/30,000,000 that of the proton, then the mass in neutrinos would be too great to have allowed the universe to have expanded in the manner observed.

the stars being two hundred thousand light-years away, and hence dim.

Aaronson and University of Arizona colleague Edward Olszewski used the Multiple Mirror Telescope to measure the velocities of stars in the Draco and Ursa Minor dwarf galaxies far more accurately than ever before. The Multiple Mirror Telescope on Mount Hopkins in Arizona is the prototype of a new generation of optical telescopes. It uses six smaller mirrors working in concert to simulate a 4.5-meter (176-inch) telescope, making it, in effect, the third-largest telescope in the world.

Aaronson and Olszewski measured the velocities of eleven stars in the Draco dwarf galaxy and ten stars in the Ursa Minor dwarf galaxy. From this snapshot they estimated the maximum velocity of stars in the galaxies. These maximum velocities are greater than the escape velocities for these galaxies if only the luminous matter is used to compute the gravity of the galaxies.

Aaronson and Olszewski's observations imply that the galaxies are roughly ten times more massive than the luminous matter would suggest; over 90 percent of the mass of these galaxies is in some dark form. This agrees with the amount of dark matter required if the dwarf galaxies are to resist the tides of the Milky Way. However, as with most methods in astronomy, Aaronson and Olszewski's method is statistical, so their findings could be a statistical fluke. The stars they measured might not be a representative sample of the stars in the galaxy. The tidal argument could also be irrelevant if the Ursa Minor and Draco dwarf galaxies are just now making their first pass by the Milky Way. In this view, they have survived not because they are more massive than was originally thought, but because they have not been previously exposed to the tidal forces of the Milky Way; we are observing these galaxies at the particular time in their existence when they happen to be passing the Milky Way for the first time. This is unlikely, but possible.

Observations of other dwarf galaxies have not clarified the situation. Pat Seitzer and Jay Frogel of the National Optical Astronomy observatories have measured the velocities of a total of fourteen carbon stars* in the dwarf galaxies in the constellations of Carina, Sculptor, and Fornax. Their observations are consistent with a dark-matter percentage of less than 30 percent in every case.

Another method used to determine the mass of a dwarf galaxy is to study its motion in a binary system. This is basically the same method used to estimate the mass of our galaxy with satellite star groups. It is subject to the same uncertainties. In a freewheeling paper, George Lake and R. A. Schommer of Bell Laboratories use a small sample of nine isolate pairs of dwarf galaxies to estimate the mass of the galaxies. They duly note that the statistical uncertainties are such that "a prudent person might just give up," but "we will press on . . ."

Lake and Schommer found that the uncertainties were too large for five of their nine pairs to say whether or not they contained dark matter, but they found evidence for massive amounts of dark matter in the other four. In two cases the dark mass was hundreds of times larger than the luminous mass. That is, over 99 percent of the mass of the galaxy is in a dark envelope!

After advising caution in the interpretation of the results, Lake and Schommer then throw caution to the winds and speculate that all dwarf galaxies have large amounts of dark matter. They go on to suggest that dwarf galaxies may be the fundamental building blocks of larger

---

*Carbon stars are bright stars believed to be in an advanced stage of evolution, as evidenced by spectra that show an unusually high abundance of carbon in their atmospheres.

galaxies. This would explain, for example, the mysterious warping or fluting that is observed to be present in the outer edges of the disks of many large spiral galaxies.

They also point out that dwarf galaxies may make their presence felt on a much larger scale. It has been known for some time that dwarf galaxies are likely to be the most numerous type of galaxy in the universe. It was believed, however, that their contribution to the total mass of universe was small, since the mass of individual dwarf galaxies was thought to be so small. If the evidence for the extremely high amounts of dark matter holds up, Lake and Schommer conclude that they may well contain most of the mass of the universe.

Perhaps God and His angels should be neglected for the noise of a fly.

CHAPTER

Hot Gas and
Dark Matter

Patience and thoroughness are the primary tools used to build a strong case. Investigators seek out all witnesses, overlook no clue, however small, and use the latest technology, whether it be a chemical analysis of a few fibers of cloth or a computer identification of fingerprints. It is unlikely that the culprit will be discovered standing over the victim with a smoking gun in hand. Nor is it likely that a dramatic courtroom confession will wrap things up neatly, as often happens in fictional mysteries.

What is needed is to construct a web of circumstantial evidence so strong that the conclusion is inescapable. The strands of this web should be independently anchored if possible, so if one strand breaks the whole web will not unravel.

Astrophysicists are faced with an analogous situation as they struggle to weave a web of evidence for dark matter. They need all the independent strands of evidence they can find. Using many different techniques, they have extended their search for dark matter to all types of galaxies.

The past twenty-five years have seen the emergence of a vigorous new branch of astronomy, X-ray astronomy. Observations with X-ray telescopes have provided an independent and powerful method for detecting the effects of dark matter.

The development of X-ray astronomy had to await the dawn of the space age. The atmosphere surrounding Earth absorbs X rays. This is fortunate for higher forms of life on Earth, but it makes life difficult for X-ray astronomers. They must use powerful rockets to put their telescopes a hundred miles or more above the surface of the planet. One of the most fruitful intellectual byproducts of the space program has been the development of X-ray astronomy. It has provided us with a new perspective of the universe by allowing us to see the hot spots, those regions of space where very-high-temperature gas exists.

In order for a gas to give off an appreciable amount of X-radiation, it must have a temperature of a million degrees or more. In the late 1950s traditional astronomers felt it was improbable that such high-temperature gas existed in detectable amounts around stars or galaxies. So, when Riccardo Giacconi and his colleagues, then at American Science & Engineering, submitted a proposal to NASA to fly a small X-ray detector on a rocket, they were turned down. The reason given was that a search for X rays from stars was unlikely to succeed.

Fortunately, Giacconi and his colleagues did not underestimate the richness of nature. They persisted. Eventually they obtained funding from the air force and launched their rockets. A new face of the universe appeared. They discovered X-ray stars, many of which were later shown to be such exotic phenomena as neutron stars and black holes. Over the years X-ray astronomers have also found hot gas around galaxies, between galaxies, and in clusters of galaxies.

As with the discovery of cold gas by radio telescopes, the discovery of hot gas by X-ray telescopes was unable to solve the dark-matter problem. There is not enough hot gas to account for dark matter. But the study of it around and in between galaxies may be of help in solving the mystery.

Hot gas is a sensitive tracer of the mass of galaxies. This is because the temperature of a gas is a measure of the average speed of the atomic particles in the gas. This speed can in turn be related to the gravity necessary to hold the particles to the galaxy.

As the atoms of a substance move faster, its temperature increases. On a very cold day ($-20°C$) the average speed of the atoms in the air is about 900 miles per hour. On a hot day ($30°C$) their average speed is 990 miles per hour. On the surface of the sun ($6,000°C$) the average speed of the atoms climbs to about 30,000 miles per hour. Atomic particles in a gas hot enough to produce X rays reach speeds of 1 million miles per hour.

These high particle speeds give direct clues as to the amount of mass a galaxy contains. The basic principle is the same as that of the daredevil horses on the carousel, except that now the horses represent not stars, but atomic particles. If the particles are moving too rapidly, they will escape from the galaxy.

This maximum speed, or escape velocity, depends on the size and mass of the galaxy. A large, massive galaxy can hold on to speedier particles better than a dwarf galaxy can. For a known galaxy size and mass, the escape velocity can be calculated. Conversely, if the average velocity of particles on the outer edges of a galaxy is known, the mass of the galaxy can be computed.

In the past few years the development of sensitive X-ray telescopes has enabled astronomers to measure the masses of certain galaxies. The galaxies contain large clouds of hot gas, and the X-ray telescopes can measure their temperature and size. From this information, astronomers can compute the mass necessary to keep the clouds from escaping into intergalactic space.

The best example of the application of this technique is the galaxy called M87. The galaxy is called M87 because it is the eighty-seventh entry of a catalog compiled by the French astronomer Charles Messier.* M87 is a supergiant elliptical galaxy. Elliptical galaxies have the shape of a basketball or football, in contrast to spiral galaxies, which are basically flat disks with spiral arms. M87 is located at the center of several thousand galaxies called the Virgo cluster. At a distance of fifty million light-years, M87 is the nearest of a class of supergiant elliptical galaxies that tend to form in the centers of clusters of galaxies. Because of its relative proximity to Earth and its unusual nature, M87 is one of the most studied galaxies.

In the late 1960s two groups of optical astronomers reported the detection of a faint optical halo around M87. This suggested that the galaxy was much larger than previously thought. A few years later X-ray observations showed that M87 is enveloped in a large cloud of hot gas.

Daniel Fabricant and Paul Gorenstein of the Smithsonian Astrophysical Observatory used data from the *Einstein* X-ray Observatory to measure the mass necessary to hold this cloud to the galaxy. They found that M87 had

---

*Messier was a comet hunter, so he made a catalog of diffuse glowing objects that might be mistaken for comets. He discovered twenty-one comets, but ironically his name is remembered not for these comets, but for his list of objects to be ignored. It includes clouds where new stars are formed, remnants of exploded stars, clusters of young stars, clusters of old stars, and many fascinating galaxies, including M31, which is the Andromeda galaxy, and M87, which is one of the largest galaxies in the universe.

Observations with X-ray telescopes indicate that the giant elliptical galaxy M87 is embedded in a massive envelope of dark matter. (National Optical Astronomy Observatories)

a mass of at least ten trillion suns and a diameter of more than half a million light-years. It is one of the most massive galaxies ever observed. Roughly 5 percent of this mass can be accounted for by stars, and the hot gas adds another 5 percent; the remaining 90 percent must be dark matter.

The X-ray observations leave little doubt that dark matter exists around M87. But M87 is obviously not a normal galaxy. It was not formed in isolation, but at the center of a rich cluster of galaxies, putting it at the bottom of a vast gravitational well. This circumstance has almost certainly enabled M87 to grow to its enormous size either at birth or later as matter torn away in collisions among other galaxies in the cluster fell to the center of the cluster.

What about normal elliptical galaxies? Do they possess envelopes of dark matter like the spiral galaxies, or are they somehow different? Certainly they look different, and close examination reveals other dissimilarities. The arms of spiral galaxies are a bright blue, indicating the presence of hot, young stars. Clouds of cool gas and dust from which new stars can form are also present. Each generation of stars returns some of its matter to the galactic gas supply, but only after a large fraction, say 50 percent, is locked up in white dwarfs, neutron stars, and black holes, never to return. As the galaxy ages, the gas supply will gradually decline; eventually it will all be gone, and the bright spiral arms traced out by newly formed stars will fade away. The central bulge will be the most prominent feature and the galaxy will appear as a disk. Elliptical galaxies, in contrast, appear to have aged prematurely. They have very little if any cool gas, and very few if any young stars.

Why are these two galaxy types so different? Did they start out that way, or is it a result of the environment in

All giant elliptical galaxies in the centers of clusters of galaxies, such as NGC 1275 in the Perseus cluster of galaxies, appear to be embedded in envelopes of dark matter.
(National Optical Astronomy Observatories)

which they evolved? It is the old argument of heredity versus environment, on a galactic scale.

Those in favor of heredity say that one or more initial, primordial qualities determined what type of galaxy a protogalactic blob would turn out to be—just as certain adherents of the heredity school might say that our genes determine whether or not we will be a genius. A quote by the poet John Dryden, "Genius must be born, and never can be taught," could be paraphrased to read: "Elliptical galaxies are born and not made through interactions with other galaxies."

The most commonly promoted "galactic gene" is rotation. The general idea is that for some reason there is an association between slow rotation and vigorous star formation when a galaxy is young. Elliptical galaxies, which rotate very slowly if at all, supposedly undergo a burst of star formation soon after they are formed. This burst blows all the gas out of the system, leaving a collection of aging stars.

Those in favor of environment argue that elliptical galaxies are made, not born. They point to the observations that show elliptical galaxies are usually found in regions of space where the density of galaxies is high, whereas spiral galaxies seem to prefer the "suburbs" of space, where the population density is low. One environmental effect that has become very popular in recent years is collisions between galaxies.

A collision between galaxies is an awesome event, but it is not as apocalyptic as one might think. It is not, for example, a vastly scaled up version of a train wreck or of a collision of an asteroid with the earth. Galaxies are not solid; they are collections of stars and gas, more like a flock of birds or a swarm of bees. They are mostly empty space. When two galaxies collide there are few direct hits between stars, just as few bees would crash if two swarms were to collide.

Several things might happen to the stars in the galaxies, depending on the relative sizes and speeds of the galaxies. On the one hand, if the galaxies were both moving at high speeds, a number of stars would be stripped from one or both galaxies, and the galaxies would go their separate ways. The stars stripped from one galaxy might be captured by the other one, or they might be left to wander through intergalactic space as "intergalactic tramps" without any galactic affiliation. On the other hand, if the galaxies were moving slowly, they might merge into one large galaxy.

The effect of a slow collision would be to puff the conglomerate galaxy into an elliptical shape. The gas clouds in these galaxies, unlike the stars, would probably collide. This would produce a burst of star formation that would effectively clear the galaxy of all cold gas. In this way, the galactic environmentalists argue, they can explain the shape of elliptical galaxies, the observed larger random velocities of the stars, the absence of cold gas, and the preference for crowded places.

Those of the heredity school counter with the question How do you explain dwarf elliptical galaxies, which show generic similarities to the giant ellipticals such as shape, lack of cold gas, and an older stellar population? It is unlikely that these systems were formed by mergers, since they are the smallest known galaxies.

Both sides agree that more information is needed about normal elliptical galaxies. Do they contain more dark matter than spirals, or less, or about the same amount?

Since elliptical galaxies rotate very slowly, this cannot be used to determine whether or not an elliptical galaxy has a dark envelope. Other optical techniques have been used to measure the average speed of stars in elliptical

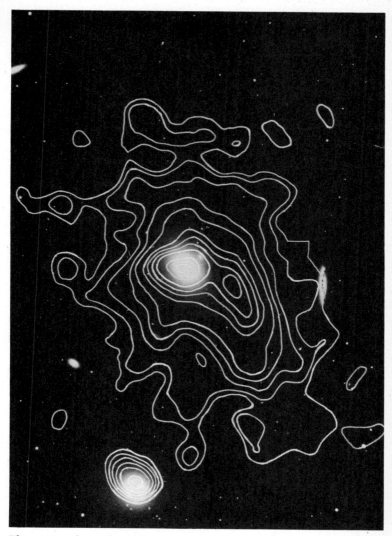

The contour lines shown here around the elliptical galaxy M86 indicate the intensity of X-ray emission, which implies the existence of a massive envelope of dark matter.
(Courtesy of W. Forman and C. Jones, Harvard-Smithsonian Center for Astrophysics; National Optical Astronomy Observatories photograph)

galaxies, but they are uncertain and are useful only for the inner parts of the galaxies. X-ray observations offer the best hope for finding dark matter in elliptical galaxies.

As the X-ray observations of galaxy clusters from the *Einstein* observatory began to accumulate, it became apparent that some more or less normal elliptical galaxies did contain gas. It is not cold gas, as in spiral galaxies, but hot gas, with temperatures of tens of millions of degrees. This was a provocative finding, because it suggested that the galaxies might need to have a massive envelope of dark matter to keep the hot gas from escaping into intergalactic space.

But there were too many uncertainties to be sure. The galaxies were all members of clusters of galaxies, so the hot-gas clouds around elliptical galaxies could be phenomena peculiar to the environment. For example, the pressure of a much larger cloud of hot gas that was known to pervade the clusters could provide a background pressure, a sort of wall around the galaxies that kept their hot-gas clouds from escaping. A survey of many elliptical galaxies not influenced by the pressure of a background gas was needed to settle the issue.

Such a survey was begun in the summer of 1983 by the husband-wife team of William Forman and Christine Jones of the Smithsonian Astrophysical Observatory. Their work showed that hot-gas clouds are a common feature of elliptical galaxies. They have substantial amounts of gas, but it is hot rather than cold. This explains why it is not observed with radio and optical telescopes.

More important, the existence of hot-gas clouds in elliptical galaxies makes it possible to estimate the mass needed to prevent the hot gas from escaping. For a typical elliptical galaxy, this mass is about five to ten times larger than the luminous mass of the galaxy. Somewhere between 50 and 90 percent of the mass of these galaxies is contained in an envelope of dark matter.

The temperature in the hot-gas cloud around the galaxies is not well determined, so the exact mass of the dark matter needed to hold the cloud to the galaxy is uncertain. But the uncertainty is such that the existence of massive amounts of dark matter in elliptical galaxies is not in question, only the exact amount is. The range given above, 50 to 90 percent of the total mass of the galaxy, takes this uncertainty into account.

X-ray observations enable us to place another strong, independent strand into the web of evidence for dark matter. There seems little room for doubt that both spiral and elliptical galaxies are embedded in envelopes of dark matter. As observational capabilities improve, new strands will be added, the existing strands anchored more securely, and the web drawn tighter.

Gravitational Lenses: Detecting Dark Matter with Bent Light

"You can hide the fire, but what you gonna do with the smoke?"
JOEL CHANDLER HARRIS

Everything has its side effects, every action its consequences. Frequently, through a careful study of these we discover what is actually happening. For example, well-concealed marijuana greenhouses are often detected through the large amounts of electricity they use.

In a similar way, envelopes of dark matter around galaxies can reveal their presence in indirect ways. A new and potentially powerful technique uses a subtle effect of dark envelopes to set a significant limit on their average size and mass. This technique is based on the predicted bending of light by a dark envelope.

One of the predictions of Einstein's general theory of relativity is that a ray of light will be bent by a gravitational field. To get a rough idea as to how this works, consider a large trampoline. If no one is on the trampoline, it is flat. If you toss a small rock onto it, there will be a small warp or dimple in it. A large rock or a person standing on the trampoline will make a large dimple. Suppose two people stand on opposite ends of it with the rocks in the middle. There will now be four dimples of differing sizes.

Now suppose that one person rolls a golf ball rapidly along the surface of the trampoline toward the other person. If the rocks and the dimples they produce are not in the way, the golf ball will roll straight. If, however, the golf ball passes close by one of the rocks, it will go off-line because of the curvature produced in the trampoline by the rock. How much the golf ball curves will depend on the mass of the rock, the speed of the golf ball, and how close the ball comes to the rock.

The gravitational bending of light works in much the same way. According to Einstein's theory, massive bodies such as stars and galaxies produce warps or dimples in space that are large or small according to whether a large or small mass is involved. A light ray passing through one of these warps is bent by an amount that depends on the severity of the warp, that is, the mass of the object, the speed of light, and how close the light ray comes to the object.

In 1911 Einstein calculated the amount and—this should gladden the heart of anyone who has suffered through a physics course, or tried to fill out an income-tax form, for that matter—he got it wrong. He estimated a deflection that was only half the correct value. The correct value includes two effects: The light wave bends because it slows down in a gravitational field around a massive object and because space is curved around the object. At the time his theory was not complete and he missed the second effect.

Five years later, he published his historic paper "The Foundation of the General Theory of Relativity." At the end of the paper he computes the bending of light correctly. A light ray just grazing the Sun is predicted to be deflected by 1.75 arc second, or roughly the angle subtended by the eye of a needle at a distance of two hundred meters—small but within the capabilities of astronomers to measure.

Because of the war in Europe, a test of Einstein's theory did not occur until 1919, when an expedition of British astronomers was mounted to observe the positions of the stars during a total eclipse of the Sun. Background stars close to the Sun in the sky can then be observed without the interference of the Sun's blinding light.

A comparison of photographs of the relative positions of the stars during the eclipse with photographs taken at another time of the year when the stars were not near the Sun showed that the light from the stars nearest the Sun had indeed been deflected. The amount of the deflection was consistent with the prediction of Einstein's theory. This triumph led to the wide acceptance of the theory and made Einstein an instant celebrity and the best-known scientist in the world.

In 1936 Einstein described how the bending of light waves by a gravitational field could magnify and brighten the images of distant stars and galaxies. In effect the star or galaxy that was bending the light rays could act like a gravitational lens. A year later Fritz Zwicky pointed out that there was a chance that galaxies would occasionally line up on the sky, so that the background galaxy would appear brighter. He proposed to measure the mass of the foreground galaxy through an analysis of the amount of brightening. This effect was beyond the capabilities of the telescopes of the day. It was forgotten until it was revived in the 1960s in an ingenious attempt to explain quasars.

Jeno Barnothy of the University of Texas suggested that quasars were not intrinsically bright; they were just moderately bright objects that appeared bright because they were magnified by the gravitational lensing of a foreground galaxy that happened to be near the line of sight to the quasar. This is not a bad idea, but it does not fit the data. As more and more quasars were discovered, more and more chance alignments were required. The gravitational-lens theory became too improbable for everyone except Barnothy, who continued to present papers on the subject to ever-dwindling audiences at astronomical meetings.

In 1979 the gravitational-lens argument was rekindled with the discovery of two identical quasars that were separated from each other by only six arc seconds. The quasars were at the same distance from Earth and the spectrum of their radiation was the same, which meant that they had the same mix of elements under the same conditions of temperature and density.

The discoverers of these objects, Dennis Walsh, Robert Carswell, and Ray Weymann of the University of Arizona, proposed that the twin quasars were really a double image of a single quasar focused by an unseen object producing a gravitational lens. Since then, additional twin quasars have been discovered by radio as well as by optical telescopes. It was hoped that these twin quasars could be used to calculate the mass of the foreground object that acts as the gravitational lens. However, this goal has proved elusive. This is mainly because it has become increasingly clear that the nature of that object is not well understood.

One difficulty is that all the quasar images come in

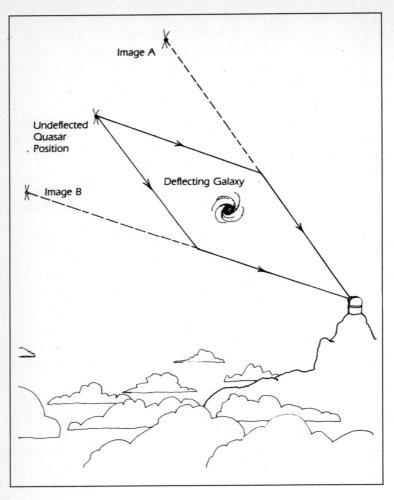

The formation of images by a foreground galaxy acting as a gravitational lens

pairs, whereas Einstein's theory as applied to the formation of images by gravitational lenses requires an odd number of images. It is possible that the third image is lost in the glare of the lensing galaxy, but unlikely that this would happen in every case. Another problem is that none of the lensing galaxies lie in a straight line with the quasar images. This means that the mass distribution of a lensing galaxy is complicated. In fact, in some cases it may not be a single galaxy. It may be that a cluster of galaxies is producing the lensing.

This possibility is supported by yet another problem. The observed separations of the twin quasars are significantly larger than expected if a single galaxy is acting as a lens. This could be because these galaxies are surrounded by other galaxies.

These uncertainties have made it impossible to derive useful information in specific cases from quasar images about the distribution of dark matter around galaxies. There is hope that statistical methods will help the situation.

Alignments of distant quasars with an intervening galaxy that are precise enough to produce images are rare— only five have been discovered so far—so a statistical treatment of these objects is not possible. While such alignments will not often produce images, they will produce a gravitational brightening of the quasar light. This may show up as a statistical enhancement of the light from distant quasars.

A more promising approach is to look for the gravitational effects on the light from distant galaxies. The light from a quasar is very compact, almost starlike in appearance. What would be the effect on light from an extended object, such as a normal galaxy? If such an object's light is bent by the gravitational field of a foreground galaxy that is slightly off the line of sight, the background galaxy image will be brightened and distorted into a kidney shape. A weaker twin image of the galaxy may also be created by the gravitational lens.

Anthony Tyson and his colleagues have undertaken an

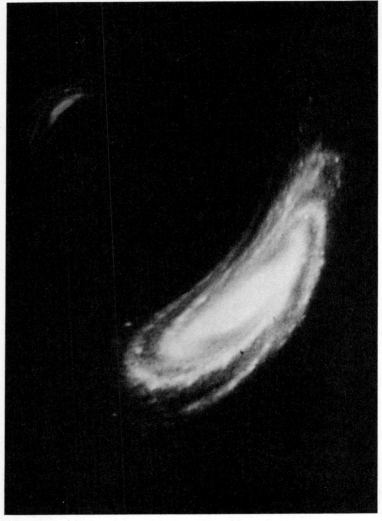

A massive object such as a foreground galaxy could warp the image of a background galaxy, as shown in this computer simulation (Copyright © 1985 Image Processing Laboratory/Smithsonian Astrophysical Observatory).

exhaustive search for kidney-shaped galaxies. In the last few years they have used the four-meter telescope on Kitt Peak to make a deep survey of selected regions of the sky. They collected a sample of almost fifty-nine thousand galaxies. These were separated into two groups, bright galaxies and faint ones.

The bright galaxies—about twelve thousand—are assumed to be in the foreground, and the faint ones—about forty-seven thousand—in the background. One possible source of error is that intrinsically faint but nearby dwarf galaxies might be mistaken for intrinsically luminous but distant normal galaxies. Using the known statistics on the number and brightness of dwarf galaxies, Tyson's group was able to show that dwarf galaxies produce only a small error, which could easily be taken into account.

The measurements of the shapes of the forty-seven thousand background galaxies showed no evidence for distortion by gravitational lensing. This allowed a limit to be set on the average mass of the foreground galaxies in their sample. More precisely, it allowed them to set a limit on the amount of mass in dark envelopes within a certain distance from the center of the galaxy.

This is analogous to setting a limit on the population of a metropolitan area. If you restrict your census to the inner city, you will obviously get a different population from the one you would get if you included the suburbs, and you would get still another number if you included outlying rural areas.

How do you determine the population of a metropolitan area? When do you stop counting? In many areas, particularly in the western United States, the population density drops rapidly outside of cities and towns, so the population of the metropolitan area levels off fairly rapidly once the suburbs have been included. For example, the number of people living within thirty miles of the center of Albuquerque is only a few percent greater than the number of people living within fifteen miles of the center.

This is the way astronomers used to believe the mass in galaxies was distributed. They thought that the density of stars and other material dropped off rapidly beyond a point roughly fifty thousand light-years from the center of the galaxy, much as the density of people drops off rapidly once you get beyond about fifteen miles from the center of Albuquerque. The mass of a galaxy could therefore be reliably estimated by measuring the mass within fifty thousand light-years of the center, just as the population of the Albuquerque metropolitan area can be reliably estimated by counting those people who live within fifteen miles of the center.

This will not work for a sprawling metropolitan area such as Los Angeles or Houston. If you try to compute the population around one of these cities, you will find that it does not level off but continues to increase after you have gone out fifteen or twenty or even thirty miles. The density of the population is less at a distance of thirty miles than fifteen miles, but not drastically less—not enough to make up for the four-fold increase in area in going from fifteen to thirty miles from the center. It is not possible in these cities to get a reliable estimate of the population of the metropolitan area by stopping at a radius of fifteen miles.

Research on galaxies in the last ten years has shown that they are not the well-concentrated shining cities we once believed them to be. Rather, they are afflicted by a peculiar kind of urban sprawl. In galaxies there is an invisible population that doesn't show up in the census of the light produced by the galaxy, yet must be included in the census of that galaxy's mass.

The population density of these invisible constituents

decreases from the center of the galaxy more slowly than the population density of the luminous matter. So slowly that the total mass of the galaxy steadily increases. It is still not clear where this sprawl stops or what the total mass of the dark matter is. All that can be done is to measure the mass inside a certain distance from the center of the galaxy.

The conventional methods use tracer stars or clouds of gas or small companion galaxies at a known distance from the center of the galaxy. Tyson's method relates to the distance of the line of sight of light rays from a background galaxy to the center of a foreground galaxy, or about three hundred thousand light-years.

The unsuccessful search by Tyson's group for kidney-shaped galaxies implies that average galaxies do not in general tend to have dark envelopes as large as the ones around our galaxy and other spiral galaxies. Why the difference? Tyson thinks that it is because most galaxies are smaller than a large spiral galaxy like the Milky Way. The statistical gravitational-lens method refers to average galaxies rather than the large spirals. In contrast, much of the work on the rotation of spiral galaxies is biased toward high-mass galaxies, because these are the ones that are easy to measure.

Does the failure to detect kidney-shaped galaxies mean that many galaxies contain very little or no dark matter? Not necessarily. The technique is not yet that sensitive. But it certainly means that the amount of dark matter is less than the extreme values derived for dwarf galaxies. There it was estimated that the dark matter outweighed the luminous matter by a hundred to one or more. The work of Tyson and his colleagues shows that a dark-matter proportion of 90 percent seems to be an upper limit for average galaxies.

In 1987 Roger Lynds of the Kitt Peak National Obser-

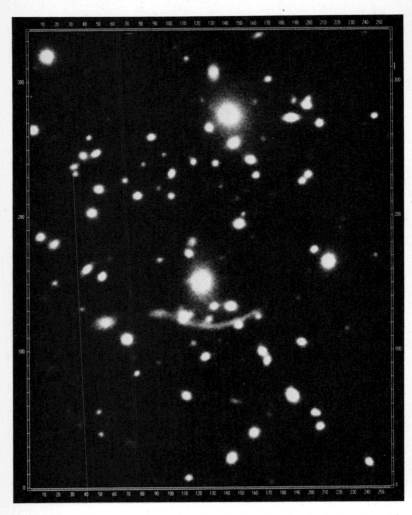

A view of Galaxy Cluster Abell 370, clearly showing the large "luminous arc." (National Optical Astronomy Observatories)

vatory and Vahe Petrosian of Stanford University found what may be the first example of a galaxy whose image has been distorted by a gravitational lens. They observed a gigantic luminous arc stretching over 300,000 light-years in the direction of a massive cluster of galaxies called Abell 370. The best interpretation of this arc is that it is the distorted image of a galaxy that is twice as far away as Abell 370. A computer analysis indicates that Abell 370 could distort the image of the galaxy into a giant arc if the cluster contains a dark matter proportion of 80 to 90 percent.

The gravitational-lensing method is still new, yet it has already shown itself to be a useful "smoke detector" that constrains the size of dark-matter envelopes. In the future, as computer-assisted surveys provide much larger samples of galaxies and the Hubble Space Telescope produces high-resolution images the method should become an increasingly important probe of dark matter in the universe.

CHAPTER
10

Dark Matter in Groups and Clusters of Galaxies

We have likened the investigation into the mystery of the dark matter to a trial. The type that it most nearly resembles is one involving corruption in government. In such a trial, the evidence usually starts at a low level, with some minor official who has taken a bribe or committed some other misdeed. The corruption is then traced to higher and higher levels until, in some cases, it turns out that practically everyone is corrupt.

So far we have established that dark matter exists in the disk of our galaxy, in an envelope or cloud around the galaxy, and in similar envelopes around galaxies of all sizes and shapes. What about the next level—groups and clusters of galaxies?

In the 1920s it became apparent, primarily through the work of Edwin Hubble at Caltech, that our Milky Way galaxy of a few hundred billion stars is but one of billions of galaxies that are spread throughout the universe as far as the most powerful telescopes can probe. Detailed surveys by Hubble and others showed that on the whole galaxies are spread uniformly around the sky. That is, there are as many in the northern sky as in the southern sky. But when studied on a finer scale, the distribution of galaxies was found not to be uniform. Galaxies clump together in groups of a few, a few dozen, a few hundred, and a few thousand. The larger groups are called clusters of galaxies. At least half of all galaxies are members of a group or cluster of galaxies.

Groups and clusters of galaxies provide an opportunity to study what might be called the "social behavior" of galaxies. This involves such questions as: Are galaxies in groups and clusters different from galaxies that evolve in isolation? How do galaxies interact? Does the presence of dark matter affect the interaction? Is there evidence for clouds of dark matter that envelop entire groups and clusters?

The answers to these questions depend on the size of the group or cluster and how closely the galaxies are packed together in a group or cluster. Our galaxy, for example, is a member of a loosely knit collection of galaxies called the Local Group.

The Local Group contains two large spiral galaxies— our Milky Way galaxy and the Andromeda galaxy, and about two dozen smaller galaxies. The latter are clustered around the Milky Way and Andromeda, which lie at opposite ends of the group, about two million light-years apart.

The most important galaxian interactions in the Local Group are between the large galaxies and their satellites. These dwarf galaxies are in orbit around the larger ones, and may eventually be pulled apart by the tides of the large galaxies. We saw earlier (Chapter 4) how a study of

The cluster of galaxies in the constellation of Hercules
(National Optical Astronomy Observatories)

the orbits of the satellite galaxies can be used to estimate the amount of dark matter around the Milky Way galaxy. We also discussed (Chapter 7) how the resistance of the dwarf galaxies to disruption by tidal forces gives one estimate of the amount of dark matter they contain.

The distance between the Milky Way and Andromeda galaxies is so great that it is doubtful these giants will ever collide or pull each other apart. They do, however, affect each other. Their mutual gravitational attraction has bound them together in a vast orbit that takes ten billion years or more to complete. At present the Andromeda is approaching the Milky Way at a speed of several hundred thousand km/hr. At this rate they will pass—probably not too closely—in about seven billion years. This knowledge of the mutual orbit of the two galaxies allows astrophysicists to estimate the combined mass of the two galaxies. It comes out to be around two or three trillion solar masses. This puts the amount of dark matter in each of the galaxies somewhere between 80 and 90 percent.

This method has been applied to other groups of two or more galaxies to give estimates of the amount of surrounding dark matter. It is, as we remarked earlier (Chapter 7), a method that must be used with caution. For example, are two galaxies that lie close together on a photographic plate part of a binary galaxy, that is, two galaxies orbiting each other, or are they the result of a chance alignment? If they are part of a binary galaxy then we can use the fact that they are moving under the influence of the gravity of their companion to determine their masses. But if the galaxies are aligned merely by chance, then the method gives a meaningless result.

Another problem is that it is impossible to know the exact position of the galaxies in their orbits. Are they slowing down to turn around in their orbit? Or are they speeding up to pass? It makes a difference in determining the mass of the galaxies. Are we viewing the orbits from the side or the top? That also makes a difference. The only recourse is to use some averaging process. If the sample is small, the averaging probably will not make sense. If it is large, it should give reliable results.

Recently, three different groups have studied over a hundred binary galaxies. Their results show that the galaxies in their samples are surrounded by massive envelopes of dark matter. These envelopes have a diameter of about half a million light-years and contain as much mass as a trillion suns. This is in agreement with the pattern established by the optical and radio measurements of the rotation of spiral galaxies, and the X-ray observations of elliptical galaxies. Galaxies are as much as ten times more massive than previously believed, and the extra mass is hidden in dark envelopes.

When several galaxies of comparable size are involved, individual galaxies move under the influence of the gravitational forces of all the galaxies in the group. In these circumstances, it is usually impossible to say anything about the amount of dark matter around one particular galaxy. But the motions of the individual galaxies can be used to compute the total amount of dark matter in the group as a whole. The technique is similar to that used for tracer stars in the galactic disk and envelope, except that tracer *galaxies* are used instead of tracer stars.

First, the velocities of as many galaxies in a group as possible are measured. Then it is assumed—this is one of the key assumptions—that the group is a stable collection of galaxies and not one that is flying apart. The observed motions of the galaxies would therefore reflect a balance between the accelerations of the galaxies and the force of gravity in the group. This is expressed by an equation that is called the virial theorem. But before the virial theorem can be used to compute the total mass of the group, another key assumption is required: that the average velocity of the galaxies in the group can be accurately calculated from the observed sample of galaxies.

Once these assumptions have been made, and the measurements taken, the total mass of the group can be computed and compared with the observed mass in stars and other luminous material. The result, based on several independent investigations: Between 70 and 98 percent of the mass in groups of galaxies is dark matter.

The upper limit has been challenged by skeptics who maintain that the case for dark matter in the universe has been overstated. Their criticism is directed primarily at the evidence from groups and clusters of galaxies. It is important to realize, however, that they do not question the existence of some dark matter around galaxies, only how much. They argue that the amount of mass in the form of dark matter is closer to 50 percent than 90 percent. The most vocal skeptics are Geoffrey Burbidge, Gene Byrd of the University of Alabama, and Mauri Valtonen of Turku University in Finland.

Although Burbidge, together with his wife, Margaret, encouraged and collaborated with Vera Rubin in her early research on the rotation of spiral galaxies, he has maintained for years that only a modest amount of matter is in the form of dark matter. He now admits that the work on spiral galaxies is convincing, but is quick to point out that this work only proves the existence of a small amount of dark matter, 50 percent or less. He has consistently taken the position that the evidence from groups and clusters of galaxies that dark matter outweighs luminous matter by ten to one in the universe is weak. He cites the difficulties of limited statistics, uncertain knowledge of the orbits of the galaxies, and the assumption that the group is stable.

"If you use the virial theorem, you undoubtedly find large amounts of dark matter," Burbidge agreed. "But, there are data on small groups that show that they are expanding, so they evidently are not stable." He was referring to the recent work of Byrd and Valtonen.

Byrd and Valtonen analyzed the data in a catalog of small groups of galaxies compiled by John Huchra and Margaret Geller of the Harvard-Smithsonian Center for Astrophysics. They found that on the average, significantly more of the galaxies seemed to be moving away from the center of the group than toward it. This is not what you would expect if the galaxies are in equilibrium. They should be swarming around the center of the group, with roughly as many moving toward the center as away from it. Byrd and Valtonen interpreted the excess numbers of galaxies moving away from the center as evidence that the groups are not in equilibrium, but are expanding. They concluded that "the need for the hypothetical missing mass is removed."

The reason the galaxies in small groups are expanding away from the group, Byrd and Valtonen suggested, is that lightweight galaxies in the groups have been accelerated beyond the escape velocity by the gravitational slingshot effect. In this process, a light galaxy picks up speed as it falls toward a more massive galaxy, but narrowly misses it. After several such near misses, the galaxy will be moving fast enough to escape from the group.

Most astronomers who are working on the dark-matter problem disagree with Burbidge, Valtonen, and Byrd. They believe that the virial theorem should provide a reasonable estimate of the masses of groups of galaxies.

"Their [Byrd and Valtonen's] work can be explained by known statistical biases," Geller asserts. For example, the evidence that the groups are expanding exists only for groups that contain a large spiral galaxy in their center.

Byrd and Valtonen were aware of this and noted without further comment that "the difference between the spiral groups and E/SO groups [elliptical galaxies and disk-shaped galaxies without spiral arms] is puzzling." To Geller and Huchra it is not puzzling. It is the key to understanding how Byrd and Valtonen reached the conclusion they did.

"Spiral galaxies are brighter," Geller explained. "So you see groups with spirals at larger distances. The more distant groups are going to have more interlopers." (An interloper in this case is a galaxy that interferes in the affairs of the astronomers by appearing to belong to a group of galaxies when in fact it is only a foreground or background galaxy.) If interlopers are mistakenly included in the type of analysis performed by Byrd and Valtonen, Geller explained, "this will give just the type of effect they report." In contrast, groups without spiral galaxies are not as bright, so they are not seen at as great a distance. These groups are not expected to have any interlopers, and they show no evidence of expansion.

The conclusions drawn by Geller and Huchra: Interlopers can explain the results of Byrd and Valtonen. There is no evidence that groups are expanding. The evidence for a dark-matter proportion of about 90 percent in groups of galaxies stands.

Summarizing this controversy, it seems that everyone agrees that groups contain at least 50 percent dark matter. This can be explained in terms of envelopes of dark matter around the individual galaxies and would require no additional dark-matter component. The skeptics would say that there is *no more* than 50 percent dark matter.

Most of the researchers who have looked into the problem feel that the dark-matter proportion could be as high as 90 percent. This too can be explained in terms of dark-matter envelopes around individual galaxies, since there

are several independent lines of evidence for dark envelopes this massive around individual galaxies.

A few astrophysicists argue for a dark-matter proportion as high as 98 percent in groups of galaxies. This
would require an extra cloud of dark matter, one that is
associated with the group as a whole. The evidence for
such a cloud is weak. It appears that all the dark matter in
groups of galaxies can be explained in terms of the dark
matter around individual galaxies.

What about the larger collections of galaxies, the rich
clusters that include thousands of galaxies? Do they contain an extra cloud of dark matter? This is not a trivial
question. It will give us an important clue as to the total
amount of dark matter in the universe, and it bears on an
important issue related to the dark-matter problem: Did
galaxies form from the fragmentation of huge, cluster-size
clouds of gas—the top-down hypothesis—or did clusters
form from the clustering together of galaxies—the bottom-
up hypothesis? The answer to this question may determine the nature of the dark matter.

The study of dark matter in clusters of galaxies began
over fifty years ago, with the irrepressible Fritz Zwicky. In
1933 Zwicky studied a particularly dense cluster of galaxies in the constellation of Coma Berenices. The galaxies in
the Coma cluster have swarmed together into the shape of
a basketball. This suggests that a balance has been
achieved between the energy of the motion of the galaxies
and the gravitational energy of mutual attraction of the
galaxies.

A similar equilibrium holds in the Sun. The motions of
the particles comprising the Sun have achieved a balance
with the gravity of all the particles. Because gravity tends
to pull particles toward one another, the stable shape of
the Sun is a ball. And since the Sun has this shape, it is a

valid assumption that a balance between gravity and the motions of the particles has been achieved. This means that we can use the virial theorem to relate the mass of the Sun to the average energy of its particles. Since we know the Sun's mass from the gravitational force it exerts on the planets, we can use the virial theorem to estimate the average energy of the particles.

Conversely, Zwicky reasoned, by measuring the average energy of the motions of the galaxies in the Coma cluster, it should be possible to estimate the cluster's mass. Zwicky, incidentally, was the first person to use the virial-theorem method for estimating the mass of groups and clusters of galaxies.

What Zwicky found was that the average mass of the galaxies in the Coma cluster is much larger than the mass of luminous matter. More than 90 percent of the mass must be hidden in some dark form, he concluded. In 1937 Sinclair Smith of the Carnegie Institution of Washington reported a similar result for the Virgo cluster.

These papers by Zwicky and Smith were the first indications that the problem of dark matter was universal in scope.

With the exception of Oort and a few others, astronomers more or less ignored the work of Zwicky and Smith until the late 1960s and early 1970s. The feeling, as it was with Oort's work on dark matter in the Milky Way, was that the problem would probably go away as observational techniques improved. Either the apparent discrepancy between the gravitational mass and the luminous mass would be shown to be spurious—the result of inadequate statistics or an erroneous assumption that the clusters were stable—or the missing mass would be found in some other form, such as hot or cool gas.

In the 1950s and 1960s observations with radio telescopes made it increasingly clear that the intergalactic

The cluster of galaxies in the constellation of Coma Berenices
(National Optical Astronomy Observatories)

dark matter could not be in the form of cool gas. They did detect cool gas in and around galaxies, but not nearly enough to account for the missing mass.

In the meantime, improved observations at visible wavelengths confirmed the earlier result that clusters contain large amounts of dark matter. Nevertheless, skepticism remained. The reasons remained the same as for the groups of galaxies: The statistical sample is inadequate and the clusters may not be stable, so the virial theorem cannot be used to estimate the mass of the cluster. This is still the position of Burbidge, Byrd, and Valtonen. They maintain that the outer parts of galaxy clusters are expanding and that the clusters will eventually disperse.

Geller and Huchra do not accept the radical interpretation of Byrd and Valtonen that the clusters of galaxies are disintegrating. "Their work on clusters suffers from the same problem as for groups, poor statistics," Geller says. She does, however, acknowledge that the amount of dark matter in clusters of galaxies may have been overestimated by a large factor.

"There are lots of biases that can lead to an overestimate of the amount of dark matter," Geller explains. "Interlopers can make you think the galaxies in clusters are moving faster than they actually are, as can errors in the calibration of the detectors. Some of the clusters are really composed of a bunch of little groups [of galaxies]. The cluster as a whole is not in dynamical equilibrium."

"Or some of the galaxies might be part of a tight binary," Huchra adds. This would mean that what is perceived as random motion of a galaxy might in fact be the motion of one galaxy around another."

"All these effects lead to an overestimate of the mass to light ratio [that is, the amount of dark matter]," Geller explains. "The actual value could turn out to be close to that of spiral galaxies." The amount of dark matter in spiral galaxies is estimated to range from 50 to 90 percent of the total.

The uncertainties inherent in the use of the virial theorem based on optical observations illustrates the need for a new method to estimate the mass in clusters of galaxies. The development of X-ray astronomy has provided such a method.

One of the early speculations was that dark matter in clusters of galaxies was in the form of hot gas, which would not show up in either radio or optical telescopes. As X-ray astronomy developed, this explanation had to be discarded. Rich clusters of galaxies were found to be pervaded with a tenuous gas that has been heated to temperatures up to 100 million degrees Celsius. The mass of this gas is considerable—it is comparable to the mass of the stars in galaxies—but it is not enough to account for the dark matter.

While the X-ray observations did not discover enough hot gas to account for the dark matter in clusters, it did provide a new method for studying the dark matter. The Einstein X-ray Observatory has shown that hot gas is a common feature of galaxy clusters. There are four possible explanations for these hot-gas clouds.

The first is that the gas is flowing out of the cluster, but is continually replenished. This would require, however, an exorbitant amount of mass and energy to maintain the flow of hot gas. Each galaxy would have to produce a hundred times its own mass in gas during its lifetime. This would literally mean that matter had to be created out of nothing, a possibility that few, if any, astrophysicists are willing to invoke. For this reason, this explanation has been rejected.

The second explanation is that we are observing the clusters at a special time in their history, the short time before the gas cloud has had time to disperse. This is improbable, given that many clusters are observed to contain hot clouds. Some are very distant and some are relatively close. The X-rays produced by the clusters have taken billions of years to reach us from distant clusters and only a hundred million years from nearby clusters. These observations from widely spread times in the histories of the clusters seems to eliminate the possibility that we are observing clusters at a special time.

The third possibility is that the hot gas is confined to the cluster by some nongravitational force. Electromagnetic forces are insufficient, since they would produce other effects that have not been observed. The pressure of an even hotter gas outside the cluster could hold the gas in the cluster, but not for long. Soon the conduction of heat from the hotter gas into the cluster gas would cause the gas in the cluster to evaporate out of the cluster.

This leaves the fourth theory. The hot gas must be confined by the gravity of the cluster. Otherwise it would expand out into the voids between the clusters. This is an important result. It means that X-ray observations can be used to measure the gravity of a cluster. From the distribution of the X-ray brightness of the gas cloud, one can estimate the distribution of the gas in space. From that, the mass needed for gravitational confinement can be estimated.

The X-ray method is not subject to the uncertainties of the optical observations that use the average velocities of galaxies. The virial theorem can confidently be used with the X-ray observations. In the case of the X-ray observations, it is the average velocity of the particles that matters. Because there are so many particles in the cloud, the averaging process can be used without any problem. And the arguments given above show that the pressure of the gas must be balanced by the gravity of the cluster, so there is no uncertainty in the application of the equations of the balance of forces.

The only significant uncertainty is that the temperature of the gas clouds cannot in every case be determined with high precision. This problem, which should be significantly reduced when more powerful X-ray telescopes are put in orbit in the next few years, is reflected in the determination of the mass (see Chapter 8).

Taking this uncertainty into account, X-ray observations indicate that the amount of dark matter required to keep the hot gas in clusters of galaxies is about five to twenty times the amount of matter in stars. But in clusters, the mass of the hot gas is comparable to the mass in stars. Apparently enormous quantities of gas were blown out of galaxies in their violent formative years. This gas is not seen around isolated galaxies because it has dispersed into the space *between galaxies* where it is trapped by gravity.

If we take into account the mass of the hot gas as an extra component of luminous matter, then the dark matter in the clusters is between about 70 and 90 percent of the total. This is consistent with the proportion of dark matter in galaxies, and is strong evidence that there is no excess component of dark matter associated with clusters.

This evidence supports the bottom-up hypothesis for galaxy and cluster formation. If galaxies had formed from a cluster-size cloud, an extra cloud of dark matter might be expected to envelop the cluster.

X-ray observations support the bottom-up scheme in another way. Images made with the *Einstein* observatory show that the clusters fall into three categories: early, intermediate, and evolved. Early systems show a loose, irregular, clumpy distribution of galaxies and a corresponding irregular and patchy gas distribution. Intermediate systems have just two or three large clumps of galaxies and gas. The

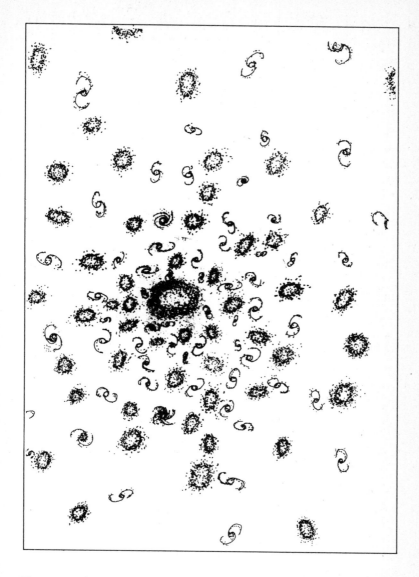

The space between galaxies in clusters of galaxies contains large amounts of hot gas trapped by the gravity of dark matter in the cluster.

evolved systems have a high central density of galaxies and
gas that falls off smoothly from the center. This sequence is
consistent with the bottom-up scheme. Randomly moving
galaxies would gradually come to equilibrium with each
other at a rate that would depend on how close together
they were initially and how rapidly they were moving. At
first they would be spread out in many clumps, but gravitational forces would gradually cause the clumps to merge
into two, and finally one, large clump.

The investigation into the nature of the dark matter in
groups and clusters of galaxies is making progress. The
proportion of dark to luminous matter seems to be the
same on this level as on the level of galaxies. The distribution of dark matter seems to be consistent with the bottom-up scheme of the formation of galaxies and clusters of
galaxies. But before we draw any firm conclusions we
should carry our investigation to even higher levels. The
next level is the realm of the superclusters.

CHAPTER

The Search for Dark Matter in the Local Supercluster

There is no good evidence for additional dark matter in the intergalactic space of groups and clusters. Does this mean that all the dark matter in the universe has been gathered into envelopes around galaxies? Is there no dark matter in the spaces outside groups and clusters of galaxies?

As the search for dark matter moves beyond groups and clusters of galaxies, the data become sparse and their interpretation becomes ambiguous. In spite of these frustrations, astrophysicists press their search, because the potential rewards are large. Most of the volume of the universe is occupied by intergalactic spaces. If these spaces contain even a comparatively low density of dark matter, the total mass there may be greater than that in all the galaxies in the universe. If astrophysicists were to ignore intergalactic spaces, they might be ignoring most of the mass of the universe!

Three types of intergalactic space can be identified. The first type is the space between galaxies in groups and clusters of galaxies. As discussed in the previous chapter, there is no evidence for additional intergalactic dark matter here.

The second type of intergalactic space, the subject of this chapter, is the space between groups and clusters of galaxies that are part of associations of groups and clusters, called *superclusters* of galaxies. Finally, there are the intergalactic voids between superclusters of galaxies.

Clusters of galaxies are at the end of the scale of the more or less regular structures in the universe. But there are irregular structures that are larger. On the sky maps of nebulae made by John Herschel and J.L.E. Dreyer in the nineteenth century, a clustering of clusters is evident. There are more galaxies in the northern hemisphere than the southern, for example, and there appears to be an almost continuous band of galaxies stretching from the Virgo cluster to a smaller cluster of galaxies in the constellation of Ursa Major. Other examples were noted by Harlow Shapley and Zwicky. But are these large irregular structures real? Or are they merely an illusion, an artifact of projecting the three dimensions of space onto a two-dimensional map?

The only way to answer this question was to determine the distances to the clusters of galaxies, a slow and painful process that required studying the spectra of hundreds of individual galaxies to get a large enough sample. Since it typically took one night of observing to get a good spectrum of a distant galaxy, this was clearly a lifetime project that few astronomers were willing to undertake.

One who did was Gerard de Vaucouleurs. For over three decades de Vaucouleurs has been working to understand the distribution and motions of galaxies within about 100 million light-years of earth. His efforts have been rewarded with deep insights, not only about this region, but the universe at large.

One conclusion of de Vaucouleurs's work is that the Local Group, which contains the Milky Way galaxy, is part of a larger aggregate of about fifty groups and clusters of galaxies. He calls this structure the Local Supercluster. He suggests that the Local Supercluster has a shape much like an enormous spiral galaxy. That is, most of the galaxies are contained within a flat disk that has a diameter of about 100 million light-years. He identifies the Virgo cluster as the center of the Local Supercluster. The Local Group lies at the edge of the supercluster disk, or the Supergalactic Plane, as de Vaucouleurs calls it.

De Vaucouleurs sketched out the properties of the Local Supercluster over thirty years ago, and in 1961 George Abell compiled a list of seventeen examples of probable superclusters. Yet it was not until the last ten years that the concept of superclusters became generally accepted by astronomers.

The problem has been one of statistics. Here, as in the case of the work of Vera Rubin and colleagues on the rotation of spiral galaxies, the breakthrough followed hard on the heels of advances in technology. The development of sensitive electronic detectors made it possible to record the spectrum of a galaxy, and hence get the information needed to determine the distance of a galaxy in half an hour rather than ten hours.

By 1983 the existence of superclusters was given the stamp of approval by the august authority of Jan Oort, who wrote a detailed review of the subject in which he stated that "there can be no doubt about the reality of superclusters."

Oort identified nine superclusters that he considered to be well established. They have diameters that are about

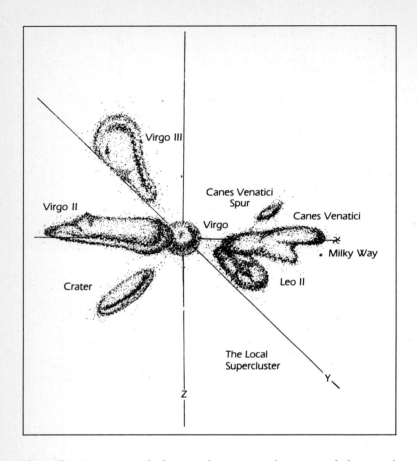

The Milky Way is part of a large agglomeration of groups and clusters of galaxies called the Local Supercluster.

ten times the size of large clusters. The most extensive superclusters contain several clusters of galaxies as well as many groups of galaxies. They generally have an irregular shape with no well-defined center—the Local Supercluster, with its pancake shape and a large cluster in its center, seems to be an exception.

*   *   *

Is there evidence for the existence of an excess of dark matter in the Local Supercluster?

This question can be addressed by considering the motions of groups of galaxies within the supercluster. For example, is the gravitational force of the supercluster pulling our Local Group of galaxies toward the center of the supercluster, and if so, by how much? If we can answer this question, then we will have an answer to the question as to how much dark matter the Local Supercluster contains.

There are two basic methods for measuring the pull of the supercluster on our galaxy. One is to study the microwave background radiation in detail. The intensity of the microwave background is uniform across the sky to less than a percent. Astronomers have continued to search for small variations in the intensity. Even a small variation can be a very large clue to a number of mysteries, such as the origin of galaxies, clusters, and superclusters, whether or not the universe is rotating, and how our galaxy moves through space.

In the late 1970s a small variation in the intensity of the microwave background was detected. Such a variation is expected from the Doppler effect if Earth is moving through the background radiation. In essence, the intensity of the radiation is amplified in the direction of motion and decreased in the opposite direction. When the orbital motion of Earth around the Sun and that of the Sun around the center of the galaxy and the galaxy within the Local Group are all taken into account, the variation in the background radiation can be used to find the motion of the Local Group. It is found to be moving in the general direction of, but not exactly toward, the center of the Local Supercluster.

Another method is to study the motions of galaxies near and far. The distant galaxies should be unaffected by the gravitational field of the Local Supercluster. Their motions are presumably due primarily to the overall expansion of the universe. They provide a background or reference against which the motions of the galaxies in and around the Local Supercluster can be compared. A systematic deviation of the motions of these galaxies from the overall expansion of the universe would be an indication that the galaxies are affected by the gravitational field of the Local Supercluster.

Over the years there has been controversy as to whether the two methods—the study of the microwave background radiation and the study of galaxy motions—give the same answer. In the most recent and detailed investigation of this type, Marc Aaronson and colleagues have concluded that they do agree. The Local Group seems to be moving with a speed of about two million km/hr in a direction that is offset by about thirty degrees from the center of the Virgo cluster.

This motion can be broken down into two parts: a motion of about one million km/hr toward the center of the Local Supercluster, and a motion of about one million km/hr of the Local Supercluster as a whole.

As a result, the motions of our planet, Earth, even when we seem to be sitting still, are quite spectacular. Earth is spinning around at about 1,600 km/hr while it orbits the Sun at about 100,000 km/hr. The Sun, in turn, is orbiting the center of the galaxy at about 800,000 km/hr. The galaxy is moving around in the Local Group at about 300,000 km/hr, and the Local Group is moving toward the center of the Local Supercluster at about 1 million km/hr. The Local Supercluster is moving, at about 1 million km/hr, in the general direction of the Hydra constellation. All this on top of the expansion of the universe, which causes distant galaxies to recede from each other at near the speed of light.

No wonder we sometimes feel breathless. And no wonder astronomers still argue about the motions of the galaxies in the universe.

The motion of the Local Group toward the center of the Local Supercluster can be used to estimate the total mass of the latter. It is assumed that this motion is due to the gravitational pull the supercluster exerts on the Local Group. The strength of this pull depends on the distribution of mass in the Local Supercluster, part of which pulls the Local Group toward the center of the supercluster and part away from the center. Our uncertainty of the exact shape of the Local Supercluster and of the position of the Local Group within the supercluster is reflected in an uncertainty in the mass estimate. Another uncertainty is the actual orbit of the Local Group. Is it falling on a direct line toward the center of the supercluster, is it making a wide circular orbit, or is it on some intermediate orbit?

When these uncertainties are taken into account, the total mass of the Local Supercluster comes out to be in the range between one and three quadrillion times the mass of the Sun. The Local Supercluster contains about one thousand large galaxies. If most of this mass is concentrated around these large galaxies, each galaxy would have to have the mass equivalent to about two trillion suns. This is slightly larger than the mass that each large galaxy is expected to have if they each have a dark halo with a radius of about a million light-years.

We can conclude that there is at present no evidence for a substantial additional component of dark matter in the intergalactic space in the Local Supercluster. But the evidence, or lack of evidence, is not strong. The uncertainties are large—they could allow for a doubling of the total amount of dark matter in the Local Supercluster—and it is not yet possible to say whether the uncertainties lead to an underestimate or overestimate of the total amount of

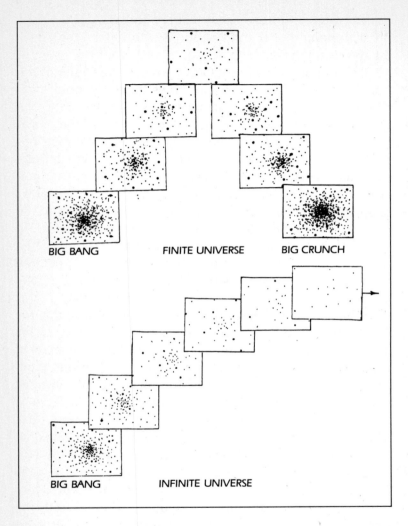

BIG BANG   FINITE UNIVERSE   BIG CRUNCH

BIG BANG   INFINITE UNIVERSE

The future of the expansion of the universe depends on whether the universe is finite or infinite.

dark matter in the Local Supercluster. A more definite conclusion must await further observational studies and the calculation of more accurate theoretical models.

It is clear that the profound issue as to whether most of the matter in the universe is hidden in intergalactic spaces cannot be answered in our local neighborhood. We must venture into the largely uncharted and even more mysterious spaces beyond the Local Supercluster.

CHAPTER

The Search for a Universal Sea of Dark Matter

As we pursue the evidence for dark matter beyond the Local Supercluster, we confront the questions that lurk in the mind of everyone who considers the mystery of the dark matter: How much dark matter does the universe as a whole contain? Does dark matter change our ideas about whether the universe is finite or infinite? Does it change our ideas about the origin and fate of the universe?

The answers to the second two questions depend on the answer to the first. According to conventional models for the universe, the finitude or infinitude of the universe is determined by the average mass density of the universe. The average mass density is the average mass in a large volume of space. It includes dark as well as luminous matter, matter in the space between galaxies, and matter in galaxies. Mass density is analogous to population density.

For example, the average population density of California is about 150 people per square mile. This number is computed by counting all the people in California and dividing by the number of square miles in California. It is an average density—the actual density is much higher than the average along the coast and much lower than the average in the desert and mountains.

The mass density of the universe plays a crucial role in the evolution of the universe. This is because it determines the strength of the gravitational force of the universe. This force is opposing the expansion of the universe in much the same way that the gravity of Earth opposes the upward motion of a rocket. If the mass density is less than or equal to a certain critical value, the gravity of the universe will be insufficient to halt its expansion. The solution of Einstein's equations for the expansion of the universe then imply that the universe is infinite and will continue to expand forever. The galaxies will move farther and farther apart, and all the stars will eventually burn out as the universe becomes cold and dark.

In contrast, if the mass density is greater than the critical density, Einstein's equations show that the universe should be finite, and that the gravity of the universe is sufficient to halt the expansion. Just as a rocket fired with insufficient thrust will eventually fall back to Earth, the universe will in this case eventually collapse. When this happens, which would be about twenty billion years in the future, the galaxies would get closer and closer together until they are crushed together in a fiery counterpart to the Big Bang that has been called the Big Crunch.

What do we know about the average mass density of the universe? Is it greater or smaller than the critical density? The mass density of luminous matter in stars and gas in galaxies and in clusters of galaxies is thought to be fairly well known. It has been estimated to be about 2 percent of the critical density. If luminous matter were all there is, then we could conclude that the universe is infinite and destined to expand forever.

But we now know that there is something in the universe besides luminous matter. There is dark matter. Is there enough dark matter to push the average density over the critical density, to tip the scales between infinity and finitude?

There are three major observational lines of evidence bearing on this question. The first is reductionist. It involves the adding up of the parts of the whole. The other two are holistic: They proceed from considerations of the evolution of the universe as a whole.

The first line of evidence consists of adding up all the matter, dark as well as luminous, in a specific region of the universe. It is then assumed that this region of space is typical of the whole universe. The calculation involves an estimate of the amount of dark matter in and around galaxies, groups and clusters of galaxies, and superclusters of galaxies.

As we have seen, these estimates are uncertain because of incomplete observations. Many different strands of evidence indicate that the relative percentage of luminous to dark matter is in the range of 50:50 to 10:90 for galaxies, groups, clusters. In the Local Supercluster, there is the possibility that enough dark matter exists in the intergalactic space between groups and clusters of galaxies to raise the overall percentage of dark matter to 95 percent. This higher figure is allowed by the uncertainties in the data. The lower figure of 50 percent is also allowed by the data.

If it is assumed that all galaxies, groups, clusters, and superclusters in the universe have this range—50:50 to 5:95 of luminous to dark matter—then the overall amount of dark matter can be estimated and the average mass density of the universe can be estimated from the average mass density of luminous matter in galaxies. This latter quantity is generally believed to be about 2 percent of the critical density that definies the boundary between a finite and infinite universe.

The assumptions made above then imply that, taking dark matter into account, the average mass density of the universe is between 5 percent and 40 percent of the critical value. The conclusion is that the universe is infinite and is destined to expand forever.

This conclusion is based on the assumptions that all the dark matter in the universe is associated with galaxies, groups, clusters, and superclusters. What if it is not? What if the intergalactic spaces between superclusters contain large amounts of dark matter?

Little is known about these intergalactic spaces. Only about 2 percent of the sky has been surveyed in the detail necessary to determine how superclusters are distributed in space and how much intergalactic space lies between superclusters. Basically what astronomers have done is take a few "deep core samples" of the sky.

The latest of these efforts is a long-term survey that is in progress under the direction of Margaret Geller and John Huchra at the Harvard-Smithsonian Center for Astrophysics. With their colleagues, they are using the sixty-inch telescope and the Multiple Mirror telescopes on Mount Hopkins in Arizona to make detailed maps of selected slices of the universe.

A description of the first slice was presented in a paper by Valerie de Lapparent, Geller, and Huchra. It covers a wedge that stretches across the sky from horizon to horizon, covers an angular width of about a dozen full moons and is three hundred million light-years deep. It contains 1,099 galaxies.

Their map suggests that the universe can be likened to a foamy bubble bath. It is mostly filled with empty spaces, or voids, corresponding to the inside of the soap bubbles. The voids range from 60 million to 150 million light-years in diameter. They contain few galaxies.

Previous, more limited surveys by other researchers had found similar voids. The work of Geller, Huchra, and de Lapparent shows that they are not anomalies. They are a common feature of the universe.

Galaxies lie along the edges of the voids, like the soap on soap bubbles. Where the voids intersect, the concentration of galaxies is increased. According to this picture the Local Supercluster has its pancake shape because it is part of a thin—20-million-light-year—shell of galaxies on the surface of a void that is 75 million or more light-years in diameter.

A second slice, recently completed, covers a region adjacent to the first slice. It confirms the bubbly structure of galaxies and voids.

Are the voids almost empty? Or are they filled with dim galaxies or gas or dark matter? If they are not empty then the estimate of the average mass density of the universe could be greatly increased, perhaps enough to tip the balance between finitude and infinitude.

Intensive searches for dim galaxies in voids are under way. The preliminary results published so far indicate that dim galaxies are distributed in space in much the same way as bright galaxies. That is, they are concentrated along the walls of the voids, not in the voids.

Recent observations by J. Ward Moody of the University of New Mexico and his colleagues have provided intriguing evidence on the contents of intergalactic voids. Their detailed survey of a large void in the constellation Boötes revealed the existence of eight previously undetected galaxies. The radiation from these galaxies is pe-

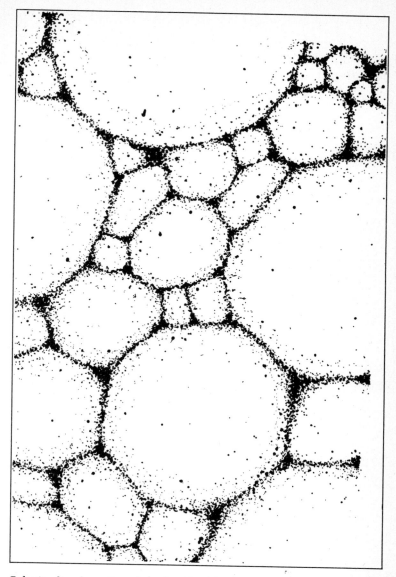

Galaxies lie along the edge of voids, like the soap on soap bubbles.

culiar; it has the characteristics of radiation from clouds of gas rather than stars. This suggests that the galaxies might be very young and full of large amounts of gas that has yet to collapse into stars, lending support to the idea that not all clumps of gas condensed into galaxies at the same time, that for one reason or another some of them may be "late bloomers." However, since eight galaxies in a region the size of the Boötes void is still only a very small percentage of the mean galaxy density in the universe, the discovery by Moody and his colleagues does not change the picture that the voids are relatively empty of galaxies.

The evidence for large amounts of gas in the voids is likewise negative. Cool-gas clouds in voids would absorb the light from galaxies lying beyond the voids in much the same way that clouds in our atmosphere can absorb sunlight. A search for this effect has been negative. There is no evidence for large amounts of cool gas in the voids.

Clouds of gas hotter than a few million degrees would not absorb the visible light from quasars. The simplest way to detect such gas would be through the X-radiation it would produce. Here an intriguing but ambiguous bit of evidence exists. Observations of wide swaths of the sky with X-ray telescopes reveal the existence of a diffuse glow of X rays that is uniform across the sky. The observed spectrum of this radiation matches very closely that expected from a gas at a temperature of several hundred million degrees. This close match is taken by a few astrophysicists as strong evidence that intergalactic space, including the voids, is filled with hot gas.

Most astrophysicists, however, doubt this conclusion. They point out the difficulties of heating so much gas to such high temperatures. Any hot gas produced in the Big Bang would have cooled well below this temperature in the first few hours of the expansion of the universe. Somehow it would have to have been reheated to several hundred million degrees. Explosive activity associated with galaxies can account for less than 1 percent of the necessary energy.

Another problem with the hot-gas interpretation is that detailed measurements of selected regions of space indicate that most of the X-radiation can be accounted for by very distant and very bright X-ray sources—presumably the explosive central regions of unusual galaxies. The apparent uniformity of the X-radiation may be an illusion, like the apparently uniform glow of lights from a distant city. There is no need to invoke the existence of clouds of intergalactic hot gas.

In summary, there is no evidence for significant amounts of dim galaxies or gas in the intergalactic voids.

Can the voids be full of dark matter? If so, galaxies would be like islands of light in a sea of dark matter. The islands might be important to those who live on them, but not in the overall balance of nature.

A uniform sea of dark matter could not be detected by any of the methods discussed so far. It would pull equally in all directions on the stars and gas in galaxies. It would be analogous, in a way, to our experience of living in the atmosphere of Earth. We normally don't notice it because it is pushing in on us equally from all directions. Yet we can detect the atmosphere. When we run or ride a horse or a bicycle or a motorcycle, we feel the ram pressure of the molecules of the air beating against our bodies and blowing our hair.

In a similar but not identical way, the motion of the Local Supercluster may give us a clue to the existence of large amounts of dark matter beyond it. If it is assumed that this motion is due to the pull of matter several hundred million light-years away, we can calculate the average density of matter in a sphere several hundred million light-years in diameter. Since such a sphere includes a number of superclusters, this might be expected to give a better estimate of the average density of the universe than an estimate based on the Local Supercluster alone.

This method yields an estimate for the average mass density of the universe in the range between 40 and 150 percent of the critical density. The difficulty with using the motion of the Local Supercluster to measure dark matter is that the motion may not be due to the gravitational pull of other superclusters. If the bubble-bath picture of the universe is correct, then superclusters are just the "soap" that has collected at the intersection of bubbles. The motion of the superclusters in this scenario might simply be a reflection of the expansion of the bubbles and not an indication of large amounts of dark matter in the voids.

Recent observations by Alan Dressler of the Mount Wilson and Las Campanas observatories and six colleagues indicate that this may be the case. They studied the motions of galaxies in a region that extends out to three hundred million light-years from our galaxy. This region includes all of the Local Supercluster, as well as several other superclusters. In particular, it includes the supercluster in the constellation of Hydra, toward which the Local Supercluster appears to be moving. Dressler and his colleagues found that all the superclusters appear to be moving as a unit. This is consistent with the picture of an expanding bubble, though, as we will discuss later, the cause of this expansion remains a profound mystery. Another possibility suggested by Dressler and colleagues is that the Local Supercluster and its neighboring superclusters are being pulled toward a much more massive supercluster. This "Great Attractor," as they call it, could contain as many as seventy-five thousand galaxies. This interpretation implies that the average mass density of the universe is between about 10 and 20 percent of the critical mass density.

The second line of evidence that might help to tie down the amount of dark matter in the universe involves another method for detecting a uniform sea of dark matter. If the sea exists, it may be possible to measure its effects on the expansion of the universe as a whole.

The galaxies in the universe can be likened to rockets fired into space. If the velocities of the rockets are great enough, they will escape from Earth's gravity. This corresponds to an infinite universe that expands forever. If the rockets' velocities are less than the escape velocity, they will fall back toward the earth. This corresponds to a finite universe with a density greater than the critical density. In either case, the velocities of the rockets decrease due to the pull of Earth's gravity.

A rocket scientist in mission control can deduce from measurements of the rate that the rockets slow down whether or not they will attain escape velocity. In a similar manner astronomers can, in principle, determine the rate at which the universal expansion is slowing down. From this information they can determine whether the universe will collapse or expand forever.

Obviously this cannot be done in exactly the same way as the rocket scientist would measure the deceleration of rockets, namely by measuring their velocities at successive times and checking to see how much they are decreasing. The expected change in the recession velocity of a distant galaxy over the last fifteen million years amounts to only about one tenth of 1 percent. This is about the level of accuracy at which astronomers can measure the velocities of galaxies, so it would take about fifteen million years to get the answer using this method.

Fortunately, other methods exist. The first is a direct method that makes use of the finite speed of light. The light observed from distant galaxies left those galaxies many millions of years ago. Therefore, we see those galaxies not as they are today, but as they were many millions of years ago. When an astronomer measures the recession velocity of a galaxy one hundred million light-years away, he is not measuring its present velocity, but the velocity it had one hundred million years ago. By comparing his measurement with the velocity the galaxy would have had if the universe had expanded without decelerating, it is possible to get a measure of the deceleration that the universe has undergone in the last one hundred million years.

The catch in this method is the "would have had" part. How can we know what velocity a galaxy would have had in the absence of deceleration? The answer is that it would have expanded at a uniform rate according to Hubble's law. This law implies that galaxies are receding from us in direct proportion to their distance.

In order to check whether Hubble's law holds for very distant galaxies, two measurements are necessary: the distance of the galaxy and its recession velocity. The recession velocity is straightforward. The red shift of the spectral lines of a galaxy can be directly interpreted in terms of a recession velocity.

Determining the distance is more complicated. Astronomers are presented with a two-dimensional picture of the universe: Stars and galaxies, glowing like distant campfires, are spread across the celestial dome. They must infer the third dimension, distance, by indirect methods. They must use clues such as the motion or the apparent brightness of the stars and galaxies.

One practical way to measure distances is the apparent brightness, or the campfire method. Imagine, for example, that you are on a vast open plain at night. Lights from many separate campfires are visible in all directions. Suppose you assumed that all the campfires had the same intrinsic brightness. Then suppose you measured the distance to a few nearby campfires by walking over to visit them. With this information you could determine the distance to more distant campfires. If a campfire appears dimmer by a certain amount, you would know that it must be farther away by an amount that can be calculated.

However, it is easy to believe that nature would never be so simple. The campfires would not all have the same intrinsic brightness. By careful observation of campfires, you might be able to identify a "standard" campfire that always has the same intrinsic brightness. In the same way, astronomers assume that certain standard galaxies have the same intrinsic brightness. As with campfires, the distance to galaxies can then be determined by comparing the brightness of these types of galaxies.

This method has been used extensively by Alan Sandage of the Space Telescope Science Institute. The "redshift program," as this project is called, has been going on ever since the two-hundred-inch Hale telescope on Mount Palomar was put into operation in 1948. By the mid-1960s Sandage was confident his measurements showed that the universe would eventually recollapse. But in 1972, after gathering much more data, Sandage wrote that "No decision is yet possible" on whether the universe will collapse or expand forever.

A major problem is the assumption that bright galaxies of a certain type have the same intrinsic brightness. There is bound to be a spread in brightness. This then leads to a spread in the calculated distances and an uncertainty in the estimate of the deceleration of the universal expansion.

An even larger source of error is the assumption that

the brightness of galaxies does not change with time. Theories of the evolution of galaxies predict that their brightness declines with time. If Sandage's results are modified to include this effect, his results then indicate that the average mass density is less than the critical density, and the universe infinite.

Similiar work by E. D. Loh and E. S. Spillar of Princeton University suggest that the issue is still in doubt. They find an average mass density between 40 and 160 percent of the critical density. This wide range of uncertainty illustrates how uncertain this method is. However, when the Hubble Space Telescope and the new generation of large ground-based optical telescopes become operational, it should be possible to reduce these uncertainties dramatically.

The third approach to estimating the amount of dark matter and the average mass density of the universe is indirect but very powerful. It involves an analysis of the elements synthesized by nuclear reactions in the first few minutes of the Big Bang—the primordial synthesis of the elements.

The exact percentages of matter that were processed into deuterium and helium in these first few minutes of the universe depend on the density and temperature during this crucial phase. Current observations of the percentage of deuterium, coupled with the Big Bang model of the universe, provide the most sensitive determination of the average mass density.

The percentage of deuterium in the universe has been estimated from its observation in molecules in the atmosphere of Jupiter, from studies of meteorites, from studies of interplanetary gas particles captured by instruments placed on the Moon, and from observations of interstellar gas. All these methods are in agreement. There is slightly more than one deuterium nucleus for every one hundred thousand hydrogen nuclei. This strongly suggests that the average mass density is in the range of 5 to 20 percent of the critical mass density, implying that the universe is infinite and will expand forever.

This is the same range of percentages obtained from the first method, namely adding up all the luminous and dark matter in galactic envelopes, clusters, and superclusters. The relative amount of dark matter in the universe implied is in the range of 50 to 90 percent.

One loophole in this argument is that the deuterium observed in the universe may not have been produced in the Big Bang. Perhaps it was produced later in supernova explosions. If this was so, then the amount of deuterium in the universe has nothing to do with the amount of dark matter in the universe and the fate of the universe. However, calculations of the amount of deuterium produced by supernova explosions or by any other mechanism give a value far below the observed amount. This loophole appears to have been closed for the present, until some ingenious astrophysicist comes up with a new mechanism.

Another, much larger loophole is that the theory of the primordial synthesis of the elements applies only to baryonic matter—normal matter composed of neutrons, protons, and electrons. However, cosmions interact only very weakly with baryonic matter. A large population of cold cosmions would have little or no effect on the percentage of deuterium manufactured in the Big Bang. The universe could contain large amounts of dark matter in the form of cold cosmions and the predicted deuterium abundance would hardly be changed.

The conclusion from the theory of the primordial synthesis of the elements must be modified to read: The average density of *baryonic* dark matter in the universe is only about 20 percent of the value needed to keep the universe from expanding forever. No limit is set on the amount of cold, cosmionic dark matter. These limits must come from the other methods described above.

Overall, the observational evidence on the amount of dark matter in the universe is not conclusive. But it is suggestive. Several independent means of estimating the amount of dark matter are consistent with the conclusion that there is about ten times more dark than luminous matter in the universe. However, there appears to be no observational evidence for a universal sea of dark matter that pervades intergalactic space and dominates the dynamics of the universe.

The implication of the observational evidence for the amount of dark matter in the universe is far-reaching. It indicates that there is not nearly enough dark matter in the universe to keep it from expanding forever.

Strangely enough, many theoretical astrophysicists disagree with the observational evidence. They argue that they have strong evidence that there is fifty times more dark matter than luminous matter in the universe! This dark matter, they say, must be in a universal sea of dark matter that is dense enough to bring the mass density up to a value that is almost precisely equal to the critical density, so the universe is balanced on a knife edge between finitude and infinity.

What is their evidence? A model of the universe, called the Inflationary Universe, that explains some of the profound problems of the standard Big Bang model of the universe so elegantly that the theorists believe it must be right. But if it is right, then about 98 percent of the matter in the universe must be in the form of a universal sea of dark matter.

What the theorists are doing is flipping the usual procedure of the scientific method on its head. Instead of using observations to support a theory, they are using a theory to say what the observations must be. In fact this is not all that unusual. It has happened before in the history of science, particularly in times when scientists were confronted with an especially puzzling set of observations or an especially compelling theory. In the next chapter we examine the Inflationary Universe model and the role it plays in the mystery of the dark matter.

CHAPTER 13

Dark Matter and the Inflationary Universe

The search for a universal sea of dark matter has put astrophysicists on the horns of a dilemma. On the one hand they can accept the observational evidence. It suggests that there is not enough dark matter to bring the total mass density of the universe up to the critical density that defines the boundary between a finite and infinite universe. If they accept this evidence, they must reject the Inflationary Universe theory, because it requires that there be enough dark matter to make the mass density of the universe equal to the critical density.

On the other hand, they can insist that the Inflationary Universe theory must be correct because it solves some difficult problems with the Big Bang theory in an elegant way. Then they must reject the observational evidence on the amount of dark matter in the universe as inadequate or incomplete in some way.

This dilemma has split the astrophysical community just about in half. There are those, mostly but not exclusively theorists, who believe that the beauty of the theory is a very strong hint that the theory is correct, that there is enough dark matter to make the mass density equal to the critical density, and that any observations to the contrary are probably wrong.

Others, mostly but not exclusively observers, believe that the best way to understand the universe is to observe it carefully and with an open mind, and to build theories on the solid foundation of observations and experiments.

At issue is much more than an academic dispute over terminology or method. The resolution of this dilemma, which is central to the mystery of dark matter, will profoundly influence our views on the origin and evolution of the universe for years to come.

The Inflationary Universe theory was developed in the early 1980s as a solution to four major problems with the Big Bang theory for the origin of the universe. With the discovery of microwave background radiation in 1965, most astrophysicists became convinced that the standard Big Bang theory, according to which the universe has evolved from a very dense hot state that existed fifteen to twenty billion years ago, was essentially the correct description for the origin of the universe. Then, in the 1970s, serious problems became apparent.

The first is the antimatter problem. Why is there so little antimatter in the universe? A particle of antimatter is a subatomic particle that has the same mass as the particle to which it corresponds but an opposite charge, among other properties. Antiparticles can be produced in high-energy collisions. These collisions occur naturally in certain cosmic settings and in particle accelerators on earth. When a particle encounters its antiparticle, they annihilate each other and their mass is converted into another form of energy, typically gamma rays.

Antimatter is conspicuous by its scarcity in our neighborhood of the universe. While tiny amounts of antimatter have been created on earth and in cosmic explosions, there is as yet no evidence for an antiplanet or antistar or antigalaxy made of antimatter. Yet, physicists, astrophysicists, and cosmologists could not come up with a good explanation as to why the Big Bang should have produced a universe that has such an imbalance of matter over antimatter.

Another problem relates to galaxy formation. It is very difficult to understand how this could have taken place in an expanding universe.

Thirdly, there is the mass-density problem. We have seen observationally that the amount of dark matter in the universe is such that the average mass density of the universe is about one fifth of the critical density. While this appears to be significantly less than the critical density, the question remains: Why is the mass density of the universe as close as it is to the critical density? Why isn't it a hundred or a million times the critical density, or a millionth of the critical density? To understand why this is a problem, consider the probability that a number drawn from a collection of numbers from 1 to 1 billion would be in the narrow range 1,000 to 5,000. It is only 4 in 1 million. In the same way, it is puzzling in the standard Big Bang theory why the mass density of the universe, which in principle could have any value, should be so close to the critical density.

Finally, there is the uniformity problem. Why is the microwave background radiation so smooth? It is uniform to less than a percent over the entire sky. This indicates that the universe must have been extremely homogeneous less than a million years after the Big Bang. The difficulty is whether or not the different parts of the universe had time to come to equilibrium.

The problem is analogous to one encountered by an

instructor in a calculus course. The answers to examination questions were remarkably uniform and correct. This suggested two possibilities: Either they were all good students with identical thought processes, or there had been some surreptitious communication between the good students and the others.

Independent evidence suggested that the latter interpretation was the correct one. The good students would finish their examinations fairly quickly, but would not turn in their papers. Instead, they would ask questions for the purpose of distracting the instructor while the answers were passed around to the rest of the class. By the time the testing period was over, the class had come to equilibrium. That is, they had uniform answers on the exam.

What bothers astronomers when they contemplate the uniformity of the microwave background is essentially the same thing that bothered the instructor. How did the universe get into equilibrium so quickly? A simple calculation based on the standard Big Bang model showed that there had not been enough time for the various parts of the universe to exchange signals so that they could all come up with the same "answer," that is, the same intensity of microwave background radiation. The radiation intensity should vary strongly from one direction of the sky to another. Yet it is remarkably uniform.

It is as if the calculus students had been separated into separate soundproof rooms so that they were out of communication, yet they had all randomly guessed the identical answers to the test. An extremely unlikely event. It is even more unlikely that all the material in the universe could have randomly come to the same temperature so as to produce the same value of the microwave background.

In 1975 Soviet astrophysicists E. B. Gliner and I. G. Dymnikova suggested that the uniformity or horizon problem could be solved simply by assuming that the universe had for some reason initially expanded very rapidly for a fraction of a second. Before the very rapid expansion the universe would have been much smaller than we would have thought by running back the presently observed expansion.

In 1980 Demosthenes Kazanas of Goddard Space Flight Center, Martin Einhorn of the University of Michigan, Katsuhiko Sato of the University of Tokyo, and Alan Guth of MIT independently proposed an explanation as to how the universe could expand extremely rapidly when it was very young. Guth, who has a flair for a turn of phrase, called his model the "Inflationary Universe," a name that was quickly picked up by other researchers.

According to the Inflationary Universe theory, when the universe was very young—a few billionths of an octillionth of a second, or 10^{-35} seconds old, it was very hot—10 octillion or 10^{28} degrees, and it was very dense. Many of the currently popular theories for the nature of interactions between particles under these conditions predict that the universe would have undergone a very brief (10^{-30} seconds) but very rapid phase of expansion, called the inflationary phase.

To see how the Inflationary Universe theory solves the uniformity problem, consider the growth of a carrot. Suppose you observed that sixty days after planting, you had six-inch-long carrots. You would conclude that the average growth rate was one tenth of an inch per day. If you assumed that carrots grew at a uniform rate, then the carrots should have been about one inch long ten days after they were planted. They were not. They were still seeds, only a tiny fraction of an inch long. The growth of carrots is not uniform. After a two-week germination period they go through a period of very rapid growth before settling down to a fairly uniform rate of growth.

Likewise, the assumption of a uniform rate of growth

could lead to a gross overestimate of the size of the early universe. The standard Big Bang theory predicts that when the universe was only 10^{-35} seconds old, all the matter in the presently observable universe—a sphere thirty billion to forty billion light-years across, was packed into a region that was slightly less than a millimeter in diameter—about the size of a carrot seed.

According to the Inflationary Universe theory, this material was packed into a region incredibly smaller than a carrot seed. The size of the region was trillions of times smaller than the radius of a proton!

So, before inflation, that part of the universe now within our horizon was much, much smaller than we would have calculated from the standard, noninflationary theory. There was no problem with communication from one side to the other, and a high degree of uniformity could have been established very early on. The observed uniformity of the microwave background is therefore a reflection of conditions that existed shortly after our part of the universe began to expand.

In the last sentence we said "our part of the universe" rather than simply "the universe" because the inflationary theory distinguishes between different domains of the universe.

In our domain we have one set of laws of physics that presumably took on their particular form when expansion began some fifteen- or twenty billion years ago. Another domain could have different laws and could have started to expand at a different time and thus have a different size. So if by the universe we mean "everything there is," then in the Inflationary Universe theory it no longer seems meaningful, strictly speaking, to talk about the age of the universe or the size of the universe, or the laws of physics that govern the universe. It only makes sense to talk about what is happening in our domain of the universe. There

SIZE

TIME

According to the Inflationary Universe theory, the universe went through a brief period of accelerated growth.

are other domains about which we shall never know.

The reason we shall never know, according to the Inflationary Universe theory, is that our domain must be enormous, many many times larger than our present fifteen- or twenty-billion-light-year horizon. The theories differ by incredible amounts on how much larger—they range from 10^{24} to 10^{1000} times larger—but it hardly matters. It is so much larger that everything we can see is piddling compared to the size of our domain. This prediction of the inflationary theory brings us back to the mystery of dark matter.

The amount of dark matter in the universe determines the overall curvature of space. This is because dark matter produces gravity and gravity curves space.

If there is enough dark matter to make the mass density of the universe greater than the critical density, then Einstein's equations predict that the mass density of the universe curves space so much that it is closed in on itself. If you sent two beams of light out along parallel straight lines in such a space, they would eventually intersect. A two-dimensional analogy is the surface of a sphere. The lines of longitude eventually intersect at the poles. This is equivalent to the statement that the universe is finite, and will eventually stop expanding and collapse.

In contrast, if the amount of dark matter is not enough to bring the mass density up to the critical density, space has a negative curvature. Parallel beams of light would not meet, but would bend farther and farther apart. Such a universe is infinite and destined to expand forever.

The dividing line between positively and negatively curved space is space with no curvature, or *flat space*. In flat space, the geometry of straight lines is familiar: Parallel beams of light remain parallel. They do not get closer together or farther apart. This would occur if the amount

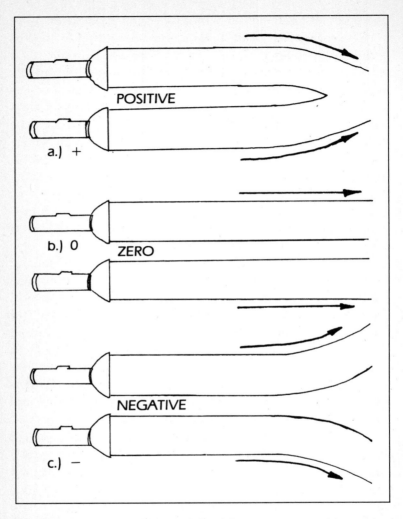

Parallel beams of light behave differently in universes with positive, zero, and negative curvature.

of dark matter is sufficient to make the mass density exactly equal to the critical density.

When astronomers try to measure the value of the mass

density, they are in effect trying to measure the curvature of space. The Inflationary Universe theory says that this is a hopeless exercise. Their measurement is bound to show that space has zero curvature, that is, that the mass density is equal to the critical density.

If you wanted to determine the shape of the earth, for example, you could get an idea that it is round by watching ships set out to sea, but you could not hope to get the same idea by observing a toy sailboat on a small pond. There you would conclude that the earth is flat. Similarly, according to the Inflationary Universe theory, astronomers are observing such a tiny part of our domain of the universe that they must conclude that space has zero curvature.

The inflationary model is an appealing model for the origin of the universe for several reasons. Probably foremost is its philosophical appeal. It uses a theory for the smallest objects in the universe—elementary particles—to explain the largest object in the universe—the universe itself.

The uniformity problem is solved in a simple, elegant way. And the mass-density question may also be resolved. The theory predicts that because of the enormous inflation in the size of the universe, we should observe space to be flat, or, equivalently, that the average mass density be extremely close to the critical density. The inflationary model also holds out considerable promise for solving the antimatter problem and the galaxy-formation problem.

In the Inflationary Universe, the dominance of matter over antimatter is attributed to elementary-particle decay processes that produce unequal amounts of matter and antimatter. These processes would have occurred when the universe was extremely hot and could have led to an excess of matter over antimatter. As the universe cooled, matter and antimatter particles annihilated each other until all of the antimatter was gone, leaving the universe with only matter.

The problem of galaxy formation may also be eased, if not resolved, in the Inflationary Universe. The "freezing out" of the various types of particles as the universe expanded could have generated clumps or concentrations of matter that acted as seeds for the formation of galaxies.

The final reason for the appeal of the inflationary model is that it makes a clear prediction: The amount of dark matter must be such as to make the mass density of the universe equal to the critical density.

That is the good news. The bad news is that, as we have seen, the observational evidence on the amount of dark matter in the universe indicates that the average mass density of the universe is only about 20 to 50 percent of the critical density. This is close, but not nearly close enough for the inflationary model, which predicts that the average mass density should be observed to be 100 percent of the critical density to an accuracy of much, much less than 1 percent.

But the appeal of the Inflationary Universe is well nigh irresistible to many theorists. "Any reasonable person has to believe that it is correct," a prominent theorist said at a recent meeting, and many scientists agreed with him. A universal sea of dark matter must exist, they maintained. The observations have just not been sensitive enough to detect it.

Not content to sit idly by and wait for better observations to be made, theorists have begun to use high-speed computers to push the search for dark matter to a deeper, more indirect level, where the predictions of computer-generated model universes are compared with observations.

CHAPTER
14

Dark Matter and Model Universes

There are two basic approaches to solving a mystery. One is to work backward from the evidence and deduce who was the culprit and how it was done. This is the way fictional mysteries are often solved. For example, in the popular mystery game of Clue it might, after a certain amount of play, be deduced that Colonel Mustard did it with the candlestick in the conservatory because there are no longer any other possibilities.

In actual mysteries, such as the mystery of the dark matter, however, the evidence is usually incomplete and the possible explanations are too varied for this approach to be successful. For example, several suspects may not have good alibis, or there may be other, unknown suspects. These cases require use of the second approach. A guess or hypothesis is made as to who did it, and a scenario is developed as to how this hypothesis can explain the evidence. Then other predictions of the scenario are checked with other lines of evidence.

Theorists are the scenarists of science. Theoretical astrophysicists construct idealized pictures or models of stars, galaxies, and the universe. They then use the laws of physics to work out the implications of these models.

If the implications of the model do not match the evidence, the theorists must go back to the drawing board. For example, the stars and clouds of gas on the outer edges of galaxies are observed to move more rapidly than the old model of the galaxy predicted. The model had to be changed, by adding an envelope of dark matter around the galaxy, and the implications worked out again. This process will be repeated until the implications match the observations well.

When future astronomical observations match the predictions of a model, the model gains credibility among astronomers and eventually becomes accepted as a reasonable explanation of the observations.

Many theoretical astrophysicists are hard at work developing scenarios that they hope will solve the mystery of dark matter. The most intense activity is in the area where the stakes are highest: the problem of the average mass density of the universe. What is the percentage of dark matter in the universe? Does the amount and exact nature of dark matter determine the fate of the universe? In particular, can the overall mosaic of groups, clusters, and superclusters of galaxies and voids be explained by one specific type and amount of dark matter and not another?

The theorists attempt to answer such questions by constructing model universes and studying their evolution. They use a large computer to follow the motion of many thousands of particles in hypothetical universes.

The standard method is to distribute the particles more or less at random in a cosmic cube that is taken to be representative of a particular model universe, and then to follow their motion under the influence of gravity and the expansion of the model universe. It is important to distinguish between random and uniform. In a uniform distribution the particles are all the same distance from one another. They form a featureless cosmic landscape—think of a vast, smooth, sandy desert.

In a random distribution, there are accidental groupings or fluctuations in which some particles are closer together than average and some farther apart—think of a vast sandy desert with barely perceptible dunes and depressions. Over the course of time, mostly through the action of wind, the sand becomes concentrated into massive dunes, the relative sizes depending on the speed of the wind, the type of sand, and so forth.

In a similar manner, the gravitational attraction between particles amplifies the random groupings or fluctuations. As time passes, more particles are pulled into these concentrations to form galaxies, clusters, and superclusters of galaxies and voids. The relative numbers and sizes of galaxies, clusters, and superclusters depends on the initial sizes of the fluctuations, how rapidly the particles were moving at first, and how rapidly the model universe expands. The approach of the model-universe builder is to choose these parameters and adjust them until the relative proportion of galaxies and the rest in the model universe is approximately equal to that in the observed universe. The particles in the model universes can be mostly baryons—ordinary protons and neutrons—or mostly cosmions—exotic invisible particles produced in the Big Bang—or a mixture. This varies from model to model.

The simplest scenario is to assume that the matter in the universe is mostly in the form of baryons since we know they exist. Most of the baryons—50 to 90 percent—would have to be in some dark form such as brown dwarfs or black holes.

Unfortunately, this simple model universe has difficulty explaining the observed mosaic of galaxies, clusters, superclusters, and voids. The problem can be traced back to the microwave background radiation. Observations of this radiation tell us that the photons were very smoothly distributed at a time as late as half a million years after the Big Bang. We also know that normal matter remained coupled to electromagnetic radiation, or photons, until the universe was about half a million years old. That is, the photons were constantly absorbed and reemitted by the matter, so they were in equilibrium; they had the same temperature.

Because of this close coupling between matter and radiation, the model-universe builders are not free to choose the initial fluctuations in a purely baryonic universe. They are severely restricted by the observed smoothness of the microwave background radiation. It implies that the baryons were very evenly distributed about half a million years after the Big Bang. How evenly? In the sand-dune analogy, it would be as if a desert roughly the size of New Mexico had no sand dunes higher than about one and a half meters, or five feet.

One way to avoid the constraints on the sizes of fluctuations that are implied by the observed smoothness of the microwave background radiation is to assume that the universe was reheated to a temperature of several thousand degrees about a hundred million years after the microwave background radiation was produced. This would break up or ionize atoms and leave electrons wandering freely through space. These electrons could have scattered the photons that constitute the microwave background radiation, thus blurring or smearing out any small-scale fluctuations produced by pregalactic clumps of matter in much the same way that fog blurs distant lights.

The reheating hypothesis represents a major departure

from the standard theory of galaxy formation. It is extremely difficult to explain how the matter in a several-hundred-million-year-old universe can be reheated from a temperature near absolute zero to several thousand degrees. One possibility is that a generation of very massive stars formed at this time. The energy from these stars could have reheated the rest of the matter in the universe. As we shall see later, this idea of a pregalactic generation of stars is central to several unconventional theories of galaxy formation. Conventional theories predict that no stars or galaxies or clusters formed until much later, when the universe had an age of a billion years or more. If we accept the conventional theories of galaxy formation, the observed smoothness of the microwave background radiation eliminates, almost completely, a universe filled with dark baryonic matter.

This difficulty in explaining the observations with a purely baryonic universe has led astrophysicists to consider model universes that are a mixture of baryons and cosmions. In such model universes, the baryons are a relatively minor component that accounts for the visible matter in the universe—somewhere between 2 and 10 percent of the total. The rest is dark matter in the form of cosmions.

Cosmions, recall, are particles that many cosmologists speculate were produced within the first fraction of a second of the expansion of the universe. If they exist, they may provide a tidy solution to the dark-matter mystery. They would interact weakly with normal matter and radiation, and they could carry most of the mass of the universe. Cosmions are sometimes called cosmic WIMPs, for Weakly Interacting Massive Particles.

Cosmions are ideal for universe builders because they lose contact with the photons very early in the expansion of the universe. This means that a clump of cosmions need not produce a clump of photons, so the observed smoothness of the microwave background radiation is not so great a problem as for universes composed purely of normal matter. The cosmions could have been highly clumped and normal matter could have gravitated toward these clumps *after* the microwave background radiation was formed. In this way, the universe builders can have their galaxies and the smoothness of the microwave background too. For these reasons, astrophysicists became very enthusiastic about cosmions in the late 1970s.

But, as one cosmologist remarked, "It's not as easy to make a universe as you might think." Even the weakly interacting cosmions must be smoothly distributed. Otherwise, they will produce regions of strong gravity that will pull normal matter into clumps too soon. The matter, in turn, will drag the photons into clumps. This would produce a patchy microwave background that is not observed. (The original version of the Inflationary Universe predicted a universe that was too clumpy. In a revised version, called the New Inflationary Universe by cosmologists, the production of clumpiness was fine-tuned to avoid this embarrassment.)

The first cosmion scenario tested extensively involved neutrinos as cosmions. Neutrinos are known to exist. They have been observed in the laboratory as a by-product of some of the same reactions that must have occurred in the early universe, and there has been some evidence, albeit controversial and unconfirmed, that neutrinos have mass.

One difficulty with neutrino models was identified by the calculations of George Blumenthal of the University of California at Santa Cruz and others. The neutrinos are hot or fast-moving cosmions. They can clump together only in large supercluster-size structures that subsequently fragment to produce clusters, then groups, then galaxies.

There is not enough clumping into groups and clusters, and it is not clear that galaxies would have formed until very recently in a neutrino universe. This is not the way the real universe looks. Groups and clusters of galaxies are common, and galaxies are at least as old as clusters and superclusters, probably older.

Detailed computer calculations of a neutrino-dominated universe by Simon White of the University of Arizona and his colleagues verified this result. Their calculations showed large filamentary structures that were reminiscent of the long chains of clusters of galaxies that had been observed. But it was too much of a good thing. They found that a neutrino-dominated universe produced too many large filamentary structures and that galaxies did not form soon enough.

Another difficulty is that dark matter in the form of neutrinos cannot be trapped around dwarf galaxies. Yet observational evidence indicates that dwarf galaxies may have dark envelopes.

The problems with neutrino universes arise because neutrinos are hot. They move too rapidly to form galaxies efficiently or to be trapped by dwarf galaxies. Cold particles would be much better for this purpose. For this reason, model-universe makers have turned to universes dominated by cold-dark-matter cosmions such as axions and photinos.

Several independent calculations have shown that cold dark matter clumps efficiently into galaxies of all sizes, from dwarfs to supergiants. These calculations also indicate that groups, clusters, and superclusters can form as well. But would the distribution of galaxies in groups, clusters, and superclusters agree with the observations?

White and his colleagues Marc Davis, George Efstathiou, and Carlos Frenk sought to answer this question. They computed the properties of cold-dark-matter

universes. They followed the motion of 32,768 particles—
each particle represents a clump of cold dark matter
slightly less massive than a galaxy. At first glance model
universes filled with cold dark matter look agreeably like
the real universe. They exhibit a variety of structures that
can be identified with groups, clusters, and superclusters
of galaxies. But a close analysis reveals problems.

The cold dark matter clumps too well. The amount of
cold dark matter in clusters and in the Local Supercluster
is predicted to be five times greater than the actual evidence implies. Observations of hot gas in clusters and the
motions of galaxies in clusters and the Local Supercluster
suggest that the amount of dark matter is about ten times
the amount of luminous matter. The model universes, in
contrast, predict that the amount of dark matter is fifty
times that of the luminous matter.

The predicted amount of dark matter can be reconciled
with the observed amount by reducing the overall mass
density of the model universes. But one of the primary
justifications for cold dark matter is that it is expected to
be produced in the inflationary universe, which requires
the higher value of the mass density.

There are other problems also. When the relative
amount of dark to luminous matter is reduced, the dark
matter becomes clumpy, like the luminous matter. This in
turn leads to a prediction that the microwave background
radiation should be clumpy. But the observed microwave
background radiation is not clumpy; it is smooth. It is extremely difficult to make a cold-dark-matter model universe that produces both a smooth microwave background
and the right mix of galaxies, clusters, and superclusters.

Finally, none of the standard model universes—with
hot or cold or baryonic dark matter—can reproduce the
bubbly structure of superclusters strung out along the sur-

face of huge spherical voids that has been observed by the group from the Harvard-Smithsonian Center for Astrophysics.

It seems that the simple model universes cannot explain the observations. Changing the amount or nature of dark matter is not enough. Something is missing, and it is almost certainly related to our understanding of how galaxies form.

In the standard model for galaxy formation, the initial fluctuations are established in the first fraction of a second of the expansion of the universe. The fluctuations are then assumed to evolve in isolation without interacting with other fluctuations. This is a stately, almost regal model of galaxy formation, according to which certain fluctuations are destined from birth to become galaxies and some are not.

This model is the one that has been traditionally used by most model-universe builders. They believe it contains the essential physics and it is clear how to make the necessary calculations. But is it a realistic model? An increasing number of theorists are having their doubts. They agree with the opinion expressed by Philip Morrison of MIT, who has said that theoretical astrophysicists working with the traditional methods "will never form galaxies. I believe galaxy-formation theory is about as good as star-formation theory was in 1925."

In 1925 quantum theory and nuclear physics were not fully developed. The crucial nuclear-fusion reactions needed to supply the energy for a star were not discovered until the late 1930s. What is missing from galaxy-formation theory, Morrison feels, "is that we don't have a picture of a complicated gravitational, hydrodynamic shock-wave process which can go in hundreds of different ways. Why should we know if we've never seen one happen? If we had

had [only] very limited data on the stars, we would never have had a stellar evolution theory. In addition, we had laboratory simulation of the nuclear reactions [for stellar evolution theory]."

Although many astrophysicists continue to use the standard model for galaxy formation, there is a growing realization that it will have to be modified to account for the observed ages, numbers, and distribution of galaxies in the universe. Two different approaches to this problem can be distinguished.

The first approach is to work backward from the observations to see what type of modification of the standard method of forming galaxies is needed to make an acceptable model universe. This is done without reference to any particular theory for the formation of galaxies. What the researchers are looking for is a mathematical prescription that describes in an acceptable way how clumps of various sizes grow into galaxies in their model universes. *Acceptable* in this context means that the prescription produces a model universe that is realistic in terms of the relative number of galaxies, clusters, superclusters, voids, and so forth. The second approach is to work from theory to observation, to develop new theories for galaxy formation, work out the consequences of these theories, and see if they match the observations.

A certain amount of success has been achieved with the first approach. A prescription called "biased galaxy formation" produces cold-dark-matter model universes that fit the observations fairly well. The computer program is written so only large clumps or fluctuations of matter become galaxies. The formation of galaxies from small clumps is arbitrarily suppressed, so a bias in favor of the large clumps is built into the model universe. This bias, if it is fine-tuned properly, alleviates one of the major problems with a cold-dark-matter universe, namely that it makes galaxies too efficiently.

Returning to the sand-dune analogy, it is as if only the large ripples, ones three times larger than the average, grow to become mountainous dunes. The rest would not grow, but would remain as low, rolling sand dunes. Most of the sand would remain spread out almost uniformly on the desert floor.

In a similar way, biased galaxy formation would leave most of the matter spread out more or less smoothly in the voids. The groups, clusters, and superclusters would no longer have too much dark matter. And since galaxies do not form as efficiently, the average distance between galaxies would be greater and clusters would be less common than in the unbiased model.

Is there any physical basis for biased galaxy formation? Several possibilities have been suggested. All of them, in ingenious ways, address Morrison's criticism that the standard model of galaxy formation is oversimplified. One is that the first galaxies to form could have influenced their environment so as to modify the formation of galaxies later. For this to work, the gas in the protogalaxies would have to have been heated to temperatures of millions of degrees. Only temperatures this high would overcome the force of gravity and prevent the protogalaxy from becoming a galaxy. This would require an enormous amount of explosive activity in the first galaxies. Such activity cannot at present be ruled out. Telescopes such as the Hubble Space Telescope may be able to detect this explosive phase of the universe, if it exists.

Another, more likely, possibility is that the bias is due to processes intrinsic to the formation of galaxies. Avishai Dekel of the Weizmann Institute of Science and Joseph Silk of the University of California at Berkeley have argued that low-density peaks in the intergalactic gas cannot form galaxies because the gas does not cool quickly enough.*

Or perhaps, when galaxies form, they go through a violent phase as the first generation of stars begin to explode. A large protogalaxy would survive this explosive phase. Its gravity would hold on to the gas left over from the formation of the massive first generation of stars. This gas would then condense mostly into many smaller, less explosive stars and the protogalaxy would settle down to become a large, bright galaxy.

A small protogalaxy, in contrast, would not have enough gravity to hold on to its reservoir of gas. Most of the gas would be blown away from the protogalaxy during the explosive phase, leaving a dim dwarf galaxy after the first generation of stars exploded. In this scheme the voids would not be empty. Rather, they would be full of dim dwarf galaxies that represented the small ripples in the primordial intergalactic gas.

A search for evidence that the voids contain an abundance of dim galaxies is under way. So far the results indicate that dim galaxies are distributed in space in much the same way as bright galaxies. There is no evidence for a preponderance of dim galaxies in voids.

Despite the lack of agreement on or observational support for any particular scheme to produce biased galaxy formation, many astrophysicists feel that it is a step in the right direction. The cold-dark-matter model universes generated on the computer look pretty much like the real universe, which makes the astronomers happy. Also, the models allow a large density, so the average mass density can be equal to the critical density as required by the inflationary Big Bang theory. This makes the cosmologists happy, for reasons discussed in the previous chapter. A

---

*Intergalactic gas cools by radiating away its energy. The radiation from the gas is proportional to the square of the density of the gas.

paper by Davis, Efstathiou, Frenk, and White concludes with a discussion of the uncertainties in their calculations, and then ends on this optimistic note: ". . . omega equal to 1 [average mass density equal to the critical density] might even be *required* to get good agreement with observation. This seems too good to be true, but perhaps it hints that we are at last approaching a correct resolution to the missing mass problem."

Other astronomers are less enthusiastic. They cite the difficulties in explaining the observed clustering of clusters of galaxies and the large-scale motions of superclusters within the framework of biased-galaxy-formation models. What is needed, everyone agrees, is a comprehensive theory for the formation of galaxies that explains biasing, the clustering of clusters, and the motion of superclusters in a natural way.

Several groups are working on new theories for forming galaxies. These theories represent radical departures from the standard theory.

Jeremiah Ostriker of Princeton University and Lennox Cowie of the University of Hawaii and, independently, Saito Ikeuchi of the University of Tokyo have proposed an explosive model for galaxy formation. In their model the galaxy-formation process resembles that of the formation of stars. Stars form from the collapse of clouds of dust and gas. Which clouds will collapse has little or nothing to do with small initial fluctuations. Rather, it is determined by the environment. Shock waves produced by stellar winds or explosions sweep through the galaxy, compress clouds, and initiate the formation of stars. Stars produce stars. In the sand-dune analogy, it is as if the formation of a large sand dune would change the wind pattern, thus facilitating the formation of another large dune.

In the standard model, galaxies form independently of

one another. In contrast, in the Ostriker-Cowie-Ikeuchi model, galaxies produce galaxies. Explosive activity in a young galaxy sends a shock wave sweeping through intergalactic space. Matter is swept up into a shell and compressed. A new galaxy is formed in this shell. Over the course of a few million years many massive stars form in the young galaxy, race through their evolutionary course, and explode. This explosive activity sends a new shock wave into intergalactic space. Matter will be swept up into a shell and another new galaxy will be formed.

As this chain reaction of galaxy formation and explosive activity develops, the shock waves from individual galaxies coalesce and form large expanding bubbles. Galaxies form along the edge of these bubbles. This would produce a pattern that is strikingly similar to the bubbly pattern that is observed. The problem is that the predicted size of the bubbles is much smaller than the observed size. Also, this model would not allow enough time for the formation of superclusters.

Despite these difficulties, astrophysicists find the explosive-galaxy-formation scenario attractive. At one time or another it has been invoked to patch up virtually every type of model universe.

In a model universe composed almost completely of baryons, that is, protons and neutrons, explosive galaxy formation is used to make galaxies form more easily. In this way it is hoped that the small initial fluctuations implied by the smooth microwave background radiation can grow rapidly enough to produce galaxies by the present day. It is unclear that this will work but it might be possible, especially if a pregalactic generation of stars reheats the universe and sets the explosive-galaxy-formation process in motion. The problem is to understand how the pregalactic stars originated. Also, it is not possible to have a universe filled with enough baryonic dark matter to make the average mass density of the universe equal to the critical density. Such a high density of baryonic dark matter would violate the constraints imposed by the primordial synthesis of deuterium and helium. But a lower density would be in conflict with the inflationary Big Bang model, which solves other problems, as discussed in Chapter 13. For these reasons, theorists have concentrated on ways to save cosmion models.

In neutrino-dominated universes the large-scale structures are not the problem. The fast-moving neutrinos naturally form clumps the size of superclusters. The problem is that they form large clumps first, and then fragment into smaller, galaxy-size clumps much later. Too late, in fact, to be consistent with the observations. These indicate that the sequence is the other way around, with galaxies forming first and then gathering together under the influence of gravity to form clusters and superclusters.

Chain reactions of explosive galaxy formation could conceivably remedy this problem. A few "seed" explosions might start a chain reaction that could quickly break up the large supercluster-size clumps into galaxy-size clumps. The formation of galaxies would then occur much earlier, as observed. The problem, again, is to understand why the original explosions that started the chain reaction occurred at just the right time and in the right manner to save the model.

The cold-dark-matter universe models, chain reactions of explosive activity in galaxies could also be useful. They could create huge expanding bubbles that would sweep baryonic matter into large shells, leaving the cold dark matter behind. This would naturally produce a bias in which the luminous matter in the galaxies does not track the dark matter. In this way it would be possible to fit most of the observations with a cold-dark-matter universe and to have average mass density of the universe equal to

the critical density, as required by the inflationary theory.
The problem is that the intergalactic voids predicted by
the theory are still too small when compared with the observed voids.

Another unorthodox theory uses an effect called
"mock gravity" to form galaxies. Mock gravity is not a
new theory of gravity. It has nothing to do with gravity. It
is concerned instead with radiation pressure.

When photons strike an object, such as a dust grain,
they exert a force on it. If equal numbers of photons hit
the dust grain from all sides, the net force on the dust
grain is zero and it will not move. An example might be a
beach ball that is blasted on all sides by streams of water.
However, if there are two beach balls side by side, they
will shield each other from the streams of water and be
pushed together. Likewise, if two dust grains lie close together, they will shield each other from part of the radiation field. An imbalance of force results and the dust
grains are pushed together, mimicking or mocking gravity.

Craig Hogan and Simon White of the University of Arizona have suggested that mock gravity may have led to the
formation of galaxies and clusters of galaxies. An attractive feature of mock gravity is that it could also sweep the
baryonic matter into large filamentary structures, leaving
behind voids filled with dark matter. Under the right conditions, it could work in either a purely baryonic, hot- or
cold-dark-matter universe.

The right conditions may not be that simple to achieve,
however. Mock gravity requires that when the universe
was a few million years old almost all the baryonic material in the universe formed into just the right combination
of stars and dust to make it work. How this would happen
remains a mystery.

The currently most popular idea for patching up model
universes is cosmic strings. Cosmic strings are a conse-

quence of some versions of the Inflationary Universe. They represent "line defects" in space-time that are in some ways analogous to line defects that occur when liquid matter crystallizes.

Strings are relics of the very early universe. Like mammoths quick-frozen in a glacier, strings were trapped when the laws of physics were "frozen" into their present form by rapidly cooling space-time. They are almost infinitesimally thin, having a typical diameter of 10^{-30} cm, much smaller than the diameter of the nucleus. Theoretical work has shown that cosmic strings, if they exist, can have no beginning or end. Either they must stretch from one end of the universe to the other, or they must have formed closed loops that have expanded to become thousands of light-years in diameter. Strings are very long and very thin. However, they have an extremely high density. A string loop that is a few thousand light-years in diameter would have a mass equal to that of a good-size galaxy.

Because of their high density, strings can produce warps in space that would bend light waves and shift the frequency of light to shorter wavelengths. If the universe was filled with enough strings to account for the dark matter, this effect would show up in the microwave background. It does not, so strings are not candidates for dark matter.

They could, however, solve some of the problems model-universe builders encounter with dark-matter candidates. For example, in a model universe dominated by massive neutrinos, galaxies do not form easily. The gravitational field around a loop of cosmic string could pull baryonic matter toward the string to form a galaxy. In this model, two types of dark matter exist. The envelopes of galaxies are composed of dark baryons, possibly brown dwarfs, or black holes, whereas intergalactic space is filled with enough massive neutrinos to make the average mass density of the universe equal to the critical density.

Advocates of cold dark matter as the solution to the dark-matter mystery also like cosmic strings. They say that large string loops could be used to form the observed long filaments of galaxies.

Ostriker, Christopher Thompson, and Edward Witten of Princeton University have proposed yet another use for cosmic strings. They argue that loops of cosmic strings have an electromagnetic field associated with them. In the early universe, these loops were vibrating violently. This vibration could generate intense electromagnetic waves that could sweep through space and clear out huge cavities. This might explain the observed intergalactic voids. Galaxies would form along the edges of the voids, producing the bubbly structure of the universe observed by the Harvard-Smithsonian group. The vibrating cosmic-string model is basically a variation of the explosive-galaxy model, with the chain reaction of galactic explosions replaced by vibrating, radiating cosmic strings.

The advantage of the vibrating strings is that they could conceivably clear out a much larger volume than the exploding galaxies. The disadvantage is that no one knows for sure that cosmic strings exist, and if they exist, whether or not they will produce the required amount of electromagnetic radiation.

Unfortunately, no convincing model universe has yet been constructed that can explain the observed mosaic of galaxies, clusters, superclusters, and voids, or the amount of dark matter that seems to be required around galaxies and in the intergalactic voids. However, a colorful galaxy of new ideas—chain reactions of exploding galaxies, mock gravity, and cosmic strings—show promise. Even if they fail to save the baryonic, neutrino, or cold-dark-matter model universes, they will have served the useful purpose of making astrophysicists rethink in more subtle and dramatic terms the ways in which galaxies are formed.

We are at the stage in the mystery of dark matter where we still have two strong candidates—neutrinos and cold dark matter—and one weak candidate—baryons. We have a rough idea of how each of them can produce the observed universe. None of the scenarios are without serious problems but all of them can be made to seem plausible with some tinkering. Astrophysicists now need to analyze these modified scenarios more carefully. Can they make more predictions? Can these predictions be observed?

In the meantime, observational astronomers need to provide more information about the bubbly structure of the universe. How large are the voids? How empty are they? Does the luminous matter track the dark matter or not?

While most astrophysical detectives revise their scenarios and search for new evidence, another, more radical possibility is being pursued as an answer to the dark-matter mystery.

CHAPTER

15

Dark Matter or a
New Law of Gravity?

Sounds emanate from the direction of a dark cellar. The normal response—that of most astrophysicists in the context of the dark-matter mystery who believe that the dark matter is there—would be to assume that something or someone is down there bumping around. Another possibility—this is the position taken by the skeptics—is that the sounds are the product of an overactive imagination, and there is nothing or no one in the cellar. There is, however, still another possibility, one that is entertained seriously by only a very few people. It is that the sounds are real, but they are not coming from the cellar. Peculiar, unknown reflections, or "sound pipes," could fool a person into thinking that something or someone was down in the cellar.

In the context of the dark-matter mystery, this is the rather lonely position taken by Mordehai Milgrom of the Weizmann Institute of Science in Rehovot, Israel. Just as we might be wrong in assuming that the normal laws governing the propagation of sound should be used to interpret the sounds from a dark cellar, Milgrom maintains that astronomers are wrong in assuming that the normal laws of gravitation should be used in interpreting observations relevant to the dark-matter mystery. In a series of articles published in 1983 he suggested that the theory used to calculate the gravitational mass should be modified.

Milgrom established a solid reputation in astrophysics with his work on the theory of X-ray stars. Then, like many of his colleagues, he became fascinated with the mystery of dark matter.

"I remember very clearly that I was struck by flat rotation curves," he said in an interview. "I felt strongly that the flat rotation curves are telling us something."

He was referring to the observation that the rotation velocity of spiral galaxies is observed to be nearly constant for wide ranges of distances from the centers of galaxies. Most astrophysicists feel the flat rotation curves are telling us that galaxies are at least twice as massive as we used to think they were, and that this extra mass is hidden as dark matter.

Milgrom took another approach. He tried to change the laws of physics.

"This idea was cooking on a very low fire," he said. "I didn't really work on it; I just had it in mind for a year or two." Then in 1980 he visited the Institute for Advanced Study in Princeton, New Jersey, for a year. John Bahcall would be there, and other astrophysicists interested in dark matter and the dynamics of galaxies would be nearby at Princeton University. Milgrom decided to devote the year to studying the dark-matter problem. He began by questioning the assumptions that underly the theory used to calculate the masses of galaxies.

What are the assumptions that underly that theory? First of all, it is assumed that the force governing the motion of atomic particles or stars or galaxies is gravity.

Secondly, it is assumed that Newton's second law of motion holds. This states that force equals mass times acceleration. For example, if you wish to have a Mack truck accelerate as rapidly as a Maserati sports car, then you must equip the truck with a very powerful engine that can deliver a large force, since the truck has so much more mass than the sports car.

The third assumption is that the gravitational force on a particle or star or galaxy is described by the universal theory of gravitation.

No one, including Milgrom, seriously questions the first assumption, that gravity is the force governing the motions involved in the dark-matter problem. Nuclear forces act only over distances on the order of the size of the nucleus, so they cannot possibly be important over galactic distances.

Electrical forces can in principle act over large distances, but in practice they do not. A strong electric field quickly attracts charged particles that set up an equal and opposite electric field that, in turn, effectively cancels the original field. It's as if you turned on a light and the light inevitably attracted so many moths that the light was blocked out past a few feet. In the same way electric fields in space are quickly canceled so they cannot affect the motions of stars and galaxies.

Magnetic forces are not as easily shielded, but it is difficult to imagine how magnetic fields strong enough to affect the motions of stars and galaxies could be generated, and if they were, they would produce other bizarre effects that have not been observed.

The only remaining force is gravity or something else. Something else is always possible and may in fact be the solution to the dark-matter mystery, but no one has thought of it yet, so that leaves gravity.

In his first articles Milgrom considered the possibility that either Newton's second law of motion or the Newton-Einstein law of gravity could be wrong. Subsequent work has shown that assuming the second law of motion is wrong leads to severe difficulties, so in later papers, with Jacob Bekenstein of Ben Gurion University, he has focused on modifications of the law of gravity.

This is the law Isaac Newton formulated to explain the motion of the Moon around Earth and of the planets and comets around the Sun. Einstein generalized Newton's law of gravitation to account for the equivalence of mass and energy and the curvature of space and time near massive objects.

The modifications introduced by Einstein's theory are essential for understanding black holes and the expansion of the universe. Newton's theory, which is now understood as an approximation to Einstein's general theory of relativity, is still considered to be adequate to explain all but the fine details of planetary and cometary motion.

Astronomers use Newton's law of gravity every day to calculate the motions of stars around each other and around the center of the galaxy and of galaxies in clusters of galaxies. Milgrom's point is that they should reconsider this everyday assumption because it may be the explanation to the mystery of the dark matter.

The standard, and by far the most popular, explanation for the mystery is that large amounts of dark matter in some yet-to-be-determined form are present in galaxies. Milgrom's unorthodox, and unpopular, hypothesis is that when the gravitational acceleration becomes very small—thirty million times smaller than the gravitational pull of the Sun on Earth—then Newton's law of gravity must be modified so that the gravitational force at large distances falls off less rapidly than prescribed by Newton's law. Specifically, he proposes that the gravitational acceleration due to an object becomes proportional to the square root of the acceleration produced by Newton's law. That is, the gravitational acceleration is proportional to the square root of the mass of the object and inversely proportional to the distance of the object.

The effect of this hypothesis is to substantially increase the gravitational force produced by a galaxy over distances of tens of thousands of light-years. The observed rapid rotation of the outer parts of spiral galaxies is no longer a problem. If the gravitational force is stronger, as Milgrom maintains it is, then dark matter is not needed to explain the excessive rotational velocities of spiral galaxies. According to Milgrom, the rotational velocity was perceived to be excessive only because astronomers have wrongly believed that Newton's theory is valid for such large distances.

Milgrom's modification of the law of gravity eliminates the need for dark matter. His modification is such that centripetal acceleration and gravity can be kept in balance only if the velocity of rotation of a galaxy tends to a constant value independent of the distance from the center of the galaxy. Thus, his hypothesis explains why the velocity of rotation for a spiral galaxy is the same at fifty thousand, a hundred thousand, and more light-years from the center of the galaxy. Milgrom also finds that, with his assumed law of gravity, the mass required to hold a rotating galaxy together depends on the velocity of rotation of the galaxy in a manner (mass proportional to the fourth power of the velocity) that is consistent with the observations.

Milgrom was initially skeptical about his own work. "I was sure that within a few hours I would find something wrong with it. I didn't, but I was sure that within a day or two I would, or a month or two. My attitude toward this in the beginning was to try to kill it."

Then, at a certain point, he changed his attitude. "I

simply went through all the possible things that I could think of to check it. I really worked. I went to bed at two in the morning and got up very early. So I don't remember very much of this period, but my wife says that she thought this would be the end of me."

He didn't mention his idea to anyone until February 1982. By then he had written up a preliminary version of it, which he sent to five colleagues. The responses ranged from lukewarm to cold, except for a fellow Israeli, Jacob Bekenstein, who is noted for his erudite contributions to such topics as the thermodynamics of black holes. Bekenstein became interested in the idea and has since co-authored papers with Milgrom on what they call "Modified Newtonian Dynamics."

Others were not so enthusiastic. "In many papers in which it would have been very natural to mention it, it was not. So, I felt it was something that people didn't want to touch. I came to visit the Institute [of Advanced Study] in September of '83 and it was sort of a taboo subject. . . . I was very surprised to the point of being naïve at their reaction. I was sure that it would be accepted much more enthusiastically. . . . When I was at Princeton I noticed people talking about a number of things that were unexplained. I felt that I came up with an explanation. I was sure that it would be [well received]. I was completely wrong. I didn't expect their reaction. . . . Since then, I have become interested in exactly how such ideas have been accepted in general, in the history of science. . . . I have learned that I should not have been surprised at their reaction."

The resistance of astrophysicists to Milgrom's hypothesis is understandable. The possibility that the Newton-Einstein theory of gravitation is wrong is just not appealing. The theory has tremendous appeal because of the simplicity of its assumptions, and because it has worked extremely well up to now. Also, astrophysicists are reluctant to abandon a theory that has served them well until they have exhausted all reasonable alternatives. The catch comes in deciding what is reasonable. After all, Newton's theory of gravity is completely adequate for most applications, yet it had to be modified by Einstein to properly explain black holes, the gravitational bending of light, the expansion of the universe, and other effects. The same could be true of Einstein's theory.

Milgrom's hypothesis has yet to achieve the rank of a serious contender to Einstein's theory. This is in part because many of the implications of his suggested modification have yet to be worked out, such as the effect on the expansion of the universe or the gravitational bending of light; it is still a hypothesis and not yet a theory. Nevertheless, his bold ideas have clearly piqued the interest of astrophysicists, and over the course of a few years there has been a perceptible shift in mood.

"I think that the attitude of people has changed a lot since the early days," Milgrom said. "Not that they believe in it. Perhaps I can describe it as having gained some respectability. You're not supposed to be ashamed to talk about it."

The best indication of the increased respect for Milgrom's hypothesis is that several astrophysicists are now trying to disprove it. In effect, this is a tribute to his work. Some of his colleagues take him seriously enough to try to prove him wrong. The extent to which they have succeeded is still unclear, but the number of his serious critics is growing.

A strong challenge has come from Lars Hernquist of the University of California at Berkeley and P. J. Quinn of Caltech and the Space Telescope Science Institute. They have analyzed the distribution of stars in the giant elliptical galaxy NGC 3923. This galaxy is observed to be surrounded by an extensive system of shells of stars, extending out to about three hundred thousand light-years. The dynamics of these shells can be understood if the galaxy is embedded in an envelope that contains about ten to twenty times as much dark matter as luminous matter. Though Milgrom has disputed their analyses, Hernquist and Quinn maintain that the shells *cannot* be explained by Milgrom's modification of the Newton-Einstein law of gravity.

Furthermore, Milgrom's theory appears unable to account for the X-ray observations of the giant elliptical galaxy M87. These indicate that large amounts of dark matter are required to explain the existence of hot gas in the inner parts of this galaxy's envelope. In this region Milgrom's form of the law of gravity agrees with the Newton-Einstein law, so he cannot account for the observations without invoking dark matter. The observations are not sufficiently accurate to be conclusive, but in this case they indicate that Milgrom's hypothesis does not solve the dark-matter problem.

Another potential problem for Milgrom's hypothesis arises from the nature of his modification of the Newton-Einstein theory of gravity. His law of gravity predicts a slower decrease of the force of gravity over large distances than does the Newton-Einstein theory. This translates into a larger discrepancy between the predictions of the Newton-Einstein law and Milgrom's hypothesis as the sizes of objects increase. Put another way, the relative amount of dark matter that must be invoked to explain the observations of clusters of galaxies, for example, must be greater than the amount of dark matter needed to explain the observations of galactic envelopes. The data, though not conclusive, do not presently support this prediction.

Finally, there is the problem of applying his modified law of gravity to the expansion of the universe. The difficulty is that the modification predicts gravitational forces that are stronger than those in the standard theory, so the expansion of the universe would be halted for mass densities much smaller than the critical density, contrary to observations.

R. H. Sanders of the Kapteyn Astronomical Institute in the Netherlands has suggested a modification of Milgrom's hypothesis that could resolve these latter two difficulties. He proposes modifying Milgrom's proposed law of gravity so that at large distances gravitational forces will once again decline in the manner of the Newton-Einstein theory. He does this by introducing a new antigravity force that acts over distances of a few hundred thousand light-years in the same manner as in the Newton-Einstein theory, as the inverse square of the distance. Inside this range, say, inside a galaxy, the antigravity force would be full strength and would oppose gravity. The net gravitational force would fall off as in the Newton-Einstein theory. Outside this range, on the outer edges of the galaxies, the antigravity force would decline and the net gravitational force would appear stronger, as in Milgrom's hypothesis. This would lead us to conclude that additional matter—dark matter—is present, when in fact it might be, as Sanders argues, that the antigravitational force is becoming weaker. A rough analogy is the motion of a kite. A kite dips toward the ground, not because gravity is suddenly stronger, but because the force of air resistance, which works against gravity, is suddenly weaker. At still larger distances, the force would again decline as in the Newton-Einstein theory, and the predicted expansion of the universe can be reconciled with the observations.

Sanders's ingenious modification might save Milgrom's hypothesis from the difficulties it encounters with clusters of galaxies and cosmological problems, but it seems to offer little hope for making it fit the data on galaxies better. New

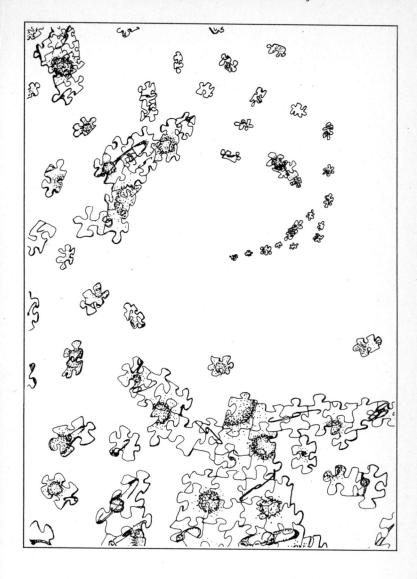

and more detailed observations will decide the issue. Even if they eventually disprove Milgrom's hypothesis, it will have served a useful purpose. By challenging one of their most cherished assumptions, Milgrom has forced astrophysicists to reexamine it carefully and to gather important new data in support of it.

An Evaluation of the Dark-Matter Mystery

Is there dark matter? If so, how is it distributed? How much of it is there? What is it? Is there more than one kind? How close are we to a solution of this mystery? We have presented evidence and arguments bearing on these questions. It is now time to evaluate the case for dark matter.

The mystery has four parts: dark matter in the disk of our galaxy; dark matter in envelopes surrounding galaxies; dark matter in groups, clusters, and superclusters of galaxies and the voids between superclusters; and the mass-density problem.

The evidence for dark matter in the galactic disk comes from studies of the motions of tracer stars. It indicates that 50 percent of the matter in the disk is in some dark form, and that this matter must be confined to the disk. This evidence, though strong, is not without its critics. They contend that an incomplete understanding of the nature of the tracer stars could lead to an overestimate of the amount of dark matter. This problem should be cleared up within the next few years.

The observations of the rotation of spiral galaxies and of X-ray emission from elliptical galaxies, together with other evidence, shows that these galaxies are surrounded by more or less spherical envelopes of dark matter. The size of these envelopes is uncertain, but they could extend as far as three hundred thousand light-years from the center of the galaxy. There is strong evidence that these galaxies, including their large envelopes, contain from five to ten times more dark than luminous matter.

The evidence for dwarf galaxies is much weaker. This is because the observations are still sparse. One of the dangers of working on the frontiers of knowledge is the temptation to draw conclusions from a limited set of observations. Those who take such risks frequently find themselves stranded far down the wrong path.

The data on dwarf galaxies could be a statistical fluke. Only a few tracer stars have been studied. Their velocities indicate that a substantial amount of dark matter is present in some galaxies but not others. Are we dealing with a real difference or inadequate statistics?

The answer to this question is crucial. If dwarf galaxies contain dark matter in similar proportions to normal galaxies, their dark matter cannot be in the form of neutrinos. Yet if we use the normally accepted criteria for assessing the uncertainty in the observations, all we can say is that the amount of dark matter in dwarf galaxies ranges from 0 to 99 percent. Clearly, the only way to answer this important question is to make more observations.

The evidence for dark matter in binary galaxies and groups of galaxies suffers from similar statistical problems. They are not so severe as for dwarf galaxies but caution should be exercised in drawing conclusions only from the data on binary galaxies and groups of galaxies.

Fortunately, we do not have to rely on these data alone. Optical and X-ray observations of galaxy clusters provide strong evidence that they contain from five to ten times as much dark matter as luminous matter. This suggests that clusters do not really pose a separate problem. It could be that all the dark matter they contain is dark matter that has been stripped by galactic collisions and tides from the envelopes around individual galaxies.

Superclusters and voids do pose a separate problem. The relative amount of dark matter in superclusters is consistent with what is found in galaxies and clusters. But it is not a simple matter to construct a model universe that has the foamy large-scale structure of the observed universe. If the galaxies formed first, as appears to have happened, then it is not clear how the superclusters and voids could have formed. In contrast, if superclusters formed first, it is not clear why galaxies appear to be so old.

The problem of explaining the large-scale structure of the universe is related to the mass-density problem. The Inflationary Universe, which elegantly explains why the microwave background radiation is so uniform, requires that the average mass density of the universe be equal to the critical density that defines the boundary between a universe that is infinite and will expand forever, and a finite universe that will eventually recollapse. Yet if all the dark matter from galaxies, clusters, and superclusters is added up, the average mass density of the universe is found to be at most one fifth of the critical density.

If enough dark matter exists to bring the average mass density of the universe up to the critical density, it must satisfy two requirements. It must be spread very uniformly throughout space, or else it would have shown up in the observations of galaxies and clusters of galaxies. Also, it cannot be in the form of normal matter, that is, neutrons and protons, or else too little deuterium would have been produced in the Big Bang.

There is also the possibility that the mass-density

problem is not a real problem. The inflationary model for the universe is an attractive one, but it may not be right. There may be another explanation for the uniformity of the microwave background radiation that does not require a period of extremely rapid expansion, or inflation. Perhaps the failure to find the amount of dark matter predicted by the inflationary model is telling us that some important physics is missing from that model.

One proposed solution to the problem of the uniformity of the microwave background radiation, other than the Inflationary Universe, is that a term has been left out of the equations that describe the evolution of the universe. This term, which is called the cosmological constant, would have no effect on the rotation of galaxies or the like. It would, however, allow for the possibility that the universe went through a "coasting" phase sometime between a few years and a few million years after the expansion began. During this phase the speed of the expansion of the universe would decrease considerably, thus allowing time for the various parts of the universe to come into equilibrium. However, no one has offered an explanation as to why the cosmological constant should have the precise value necessary to produce a coasting phase at the appropriate time.

What is the dark matter? The possible answers to this question can be grouped into three general categories.

One is that there is no dark matter, that the laws of physics are wrong. Milgrom's modification of the law of gravity is the most studied possibility in this realm.

Another category is baryonic matter—normal matter composed of neutrons and protons, but assembled in some dark form, such as low-mass brown-dwarf stars or black holes. Dark baryonic matter, if it exists, must have assumed its present form either before galaxies formed or when they were forming.

David Criswell, a space scientist who has written extensively on the industrialization of space, has proposed a provocative variation on the brown-dwarf theme. He suggests that the dark matter in galactic disks and envelopes is composed of stars and planets that are being husbanded by advanced extraterrestrial civilizations. These matter- and energy-hungry civilizations would gather all the material in their vicinity into industrial facilities, space homes, and reservoirs of hydrogen and helium gas. They would even dismantle most of their star for this purpose. Criswell speculates that technologically advanced civilizations could use facilities similar to large ion accelerators on earth to create enormous electric currents and magnetic fields in space. These magnetic fields could be used to dismantle planets by increasing their spin until they break apart, or to dismantle stars by pumping energy into their upper atmospheres. Since a low-mass star burns its energy much more slowly than a high-mass star, these civilizations could extend the lifetime of their star from about ten billion years for a solar-type star to ten quadrillion years by storing the hydrogen and helium in reservoirs for future use in fusion reactors. The best way to store this material would be in thousands of self-gravitating, planetlike objects—artificial brown dwarfs! These might be detected as a cluster of infrared sources. There may be other means of detecting such civilizations. For example, strong magnetic fields used to move large amounts of matter around or radiation from controlled thermonuclear fusion might be detectable.

Is the dark matter evidence for galactic development projects? Are we like inhabitants of some remote jungle who have climbed the top of a mountain and seen, for the first time, a vast and complex interstellar fuel refinery? It is an intriguing idea, and not that much more farfetched than some of the other explanations for dark matter.

The final category is nonbaryonic matter, or cosmions. This matter might have been produced in the very early stages of the Big Bang. It does not radiate and shows up primarily through its gravitational effects.

Some groups of cosmions are hot—fast-moving particles such as neutrinos with mass—and some cold—slowly moving particles such as photinos or axions. It is important to keep in mind that particles with the required properties are suggested by some theories but have never been observed. Nor are they the only possible types of cosmion. Different theories for the origin of the universe and elementary particles imply various types of cosmions.

The *superstring* theory is the latest vogue for the theory of everything—all particles and all forces—and though not directly related to cosmic strings, it derives its name from the assumption that fundamental particles are not points, but tiny—10^{-33} cm in diameter—loops of strings. It calls for almost everything the other theories do in the way of exotic dark-matter candidates and more. It allows for the possibility of another form of matter, called "shadow matter."

Shadow matter is distinct from all other types of matter we have considered. It is not quarks or leptons or photons or the supersymmetric partners of these particles. The only connection between the world of shadow matter and the world as we know it would be through the gravitational field of shadow matter. It is somewhat like cosmic strings, except that shadow matter is not restricted to stringlike configurations. Shadow stars, shadow mountains, and shadow people could conceivably exist, but we would be aware of them only through their gravitational force. It is akin to the notion of a parallel universe that penetrates our universe but does not interact with it. According to the superstring theory, in the earliest stages of the expansion of the universe, when all the forces were unified as one superforce, shadow matter was mixed in with ordinary matter—there was no difference between the two.

The theory maintains that the first great fragmentation of the forces of the universe occurred about 10^{-43} seconds after the beginning of the expansion. It was then that the force of gravity separated off from the other forces, and that the shadow universe separated off from the normal universe. Could shadow matter be the explanation of the dark-matter mystery? Can a test be made for the existence of shadow matter?

Michael Turner of the University of Chicago and his colleagues Edward Kolb and David Seckel of the Fermi National Accelerator Laboratory considered a number of possibilities. The first was that the shadow universe exactly mirrored the ordinary world. That is, the microphysics that governed the shadow world was an exact counterpart to our world. The mass of shadow quarks and leptons were the same as quarks and leptons, and the law of shadow electromagnetism was the same, and so forth.

Does this mean that Earth is half shadow matter, and that we are sharing our beds with shadow people? No. Suppose there was a shadow person walking around with us, and we stepped on a scale. If the scale operated with springs or electromagnetic forces, as an ordinary bathroom scale does, it would not interact with our shadow person and would not weigh him or her. If, however, the scale operated with gravitational forces, such as a balance, it would weigh the shadow person. We would weigh twice as much as on the bathroom scale. This does not happen, so we know there are no shadow people. Similar arguments can be used to eliminate the possibility that Earth as a whole contains much, if any, shadow matter.

The Sun maintains a balance between internal pressure and gravity through nuclear reactions in its interior.

Because shadow matter does not participate in such reactions, but does provide gravity, the pressure in the core of the Sun would have to be increased to support the extra gravity provided by shadow matter. But this increase of pressure would increase the luminosity, surface temperature, and neutrino output of the Sun. The requirement that the predicted properties of the Sun agree with the observed properties limits the amount of shadow matter in the Sun to less than 1 percent.

The overall amount of shadow matter in the universe is limited by the amount of helium produced in the Big Bang. The presence of more than a small percentage of shadow matter would result in an overproduction of helium. Thus an exact mirror shadow world of the same microphysics cannot possibly explain the dark-matter problem.

There is a possible, though not simple, way around this problem. If the microphysics of the shadow universe is not the same as for the normal universe, it may be possible to construct a shadow universe that does not lead to an overproduction of helium in the Big Bang. It is difficult to know how to test for the existence of such a world. Would shadow matter clump to form stars? If so, how large would the stars be? Or would it remain in clouds of shadow gas, and if so, how large would they be? Until a way can be found to test the predictions of these shadow universes with shadow microphysics they really can't be considered to constitute a scientific hypothesis because they lack the central element of such a hypothesis: testability.

What about other hypotheses? How can we decide between changing the laws of physics or baryonic dark matter or hot dark matter or cold dark matter? At this stage it is not easy. There are solid arguments that rule out baryonic dark matter *if* the overall density of the universe is equal to the critical density. But how convinced are astrophysicists and cosmologists that this must be true? How do we evaluate the validity of hypotheses?

There exists a formal method for evaluating hypotheses in terms of probabilities. It is called Bayesian statistics. Bayesian statistics, named after Thomas Bayes, an eighteenth-century British clergyman and mathematician, involves the use of probability theory to calculate the probability that a hypothesis is correct, given a set of observations; this probability is then updated as the observations are improved. For example, Bayesian statistics or related methods are used to evaluate the hypothesis that a particular straight line or curve is the best fit to a particular set of observational or experimental data points. In a similar way, the method could be used to estimate the probability of the hypothesis that brown dwarfs account for the dark matter in the disk of our galaxy, given the present set of observations. Success or failure in observing brown dwarfs in the future could then be factored into the equation to increase or decrease the probability that the brown-dwarf hypothesis is correct. Bayesian statistics is a potentially useful tool, but astrophysicists and other scientists generally use it, or related methods, only for analyzing data. They rarely apply it to hypotheses that pull together a group of observations. Perhaps they have yet to be convinced of its usefulness. Instead they use subjective criteria such as simplicity and beauty while they wait for the observational evidence to accumulate to the point where they can use the ultimate criterion, the observational test.

Unfortunately, in the search for the solution or solutions to the dark-matter mystery, we are still at an intermediate stage. It has been possible to rule out a few hypotheses. Red dwarfs and neutron stars cannot account for the dark matter on any level. Baryonic matter of any type cannot solve the mass density problem. Cosmions, hot or cold, cannot account for the dark matter in the disk of our galaxy. Nor can either hot or cold cosmions simultaneously solve the other three dark-matter problems—galaxies, superclusters, and the mass-density issue—without additional complicated assumptions such as biased galaxy formation, or explosive galaxy formation, or cosmic strings.

The table on page 215 lists the types of dark-matter problems, the candidates, and an assessment of their candidacy in terms of "Yes," "No," and "Maybe." Also listed is a noncandidate, the modification of the laws of gravity suggested by Mordehai Milgrom and Jacob Bekenstein. We have been generous with "Maybes." For example, white dwarfs seem very unlikely to solve the galactic-disk problem, neutrinos seem very unlikely to solve the galactic-envelope problem, and few scientists believe that a modification of the laws of gravity are the solution to any of the problems, but we have entered "Maybe" because there is still a possibility for success.

The entries in the table show how difficult it may be to solve the mystery of the dark matter. Except for the possibility, deemed unlikely, that the laws of gravity will have to be changed, none of the candidates can solve all the dark-matter problems. Every candidate has at least one "No" and one "Maybe."

A case in point is cosmions. Because of their rapid motion they cannot explain the dark matter in the galactic disk. Some other solution, most likely baryonic matter, must be invoked to solve this problem. Yet baryonic dark matter cannot solve the mass-density problem. So if the mass-density problem is real there must be at least two types of dark matter.

The prospects for a simple solution to the other two problems listed in the table (envelopes and superclusters)

Rating of Dark-Matter Candidates

Candidate	Problem			
	Galactic Disk	Galactic Envelopes	Super-clusters and Voids	Mass Density of Universe
Black Holes	No	Maybe	Maybe	No
White dwarfs	Maybe	No	No	No
Brown dwarfs	Maybe	Yes	Maybe	No
Hot dark matter (Neutrinos)	No	Maybe	Yes	Yes
Cold Dark Matter	No	Yes	Maybe	Yes
Modification of Gravity	Yes	Maybe	Maybe	Maybe

If a particular candidate is ruled out for a particular problem, a "No" is entered. If a candidate is not ruled out, a "Yes" is entered. If there is evidence against a candidate, but uncertainties in the data, or additional assumptions may be able to save the candidacy, a "Maybe" is listed.

are not much brighter. Cold dark matter is good for envelopes around galaxies but not superclusters. Hot dark matter is good for superclusters but not galactic envelopes.

One ingenious class of solutions to this problem is to postulate a cosmion that changes or decays from cold dark matter to baryons and hot dark matter at the appropriate time. According to this scenario, most of the matter in the universe would be cold dark matter until galaxies formed, at which time all or part of it would decay into baryons and hot dark matter so that superclusters and voids could form. This type of model requires careful *ad hoc* fine-tuning of the time of decay and the types of decay products. Observational constraints from the microwave background and the observed distribution of matter in dark envelopes may eventually eliminate such models, but they remain a possibility.

Others don't try to be so inventive. They simply state

that all types of matter—baryons, hot and cold dark matter—are produced in the Big Bang in the right amount necessary to solve the various dark-matter mysteries. This is most unappealing from the point of view of simplicity and beauty. But are the criteria of simplicity and beauty reliable guides to the truth?

A very old and illustrious tradition insists that they are. Pythagoras used them to arrive at his conclusion that Earth is round. Copernicus used them to place the Sun at the center of the solar system. Einstein used them when formulating his theory of relativity. Paul Dirac, who won a Nobel Prize in physics for work that predicted the existence of antimatter, succinctly expressed the working credo of many theoretical physicists: ". . . if one is working from the point of view of getting beauty in one's equations, and if one has sound insight, one is on a sure line of progress."

In keeping with this tradition, dark-matter theorists are searching for a simple and beautiful theory, one in which all the pieces of the puzzle fall into place. Of course, not all beautiful theories turn out to be true and not all "ugly" theories are necessarily false. Simplicity and beauty are only guides, not proofs. The ultimate criterion for a theory is testability.

Even though the search for the solution to the mystery of dark matter has not turned up any beautiful solutions, it has generated a number of hypotheses or theories that can be tested.

If brown dwarfs make up the dark matter in the disk of the galaxy, it should be possible to observe a number of them with sensitive infrared telescopes within the next few years.

The gravitational bending of light around stars should make further tests possible. Stars of different masses produce different types of effects. This method can be used to exclude the existence of black holes in numbers sufficient to make the mass density of the universe the critical density. It cannot now eliminate the possibility that they could explain the dark envelopes around galaxies, but in the next decade this too may be possible.

It may even be possible to detect cosmic strings through their gravitational effect on light. They would produce a galaxy with a sharply cut off edge. It has been estimated that one or more such objects should be detected in the first few years of operation of the Hubble Space Telescope, if cosmic strings are to explain superclusters of galaxies.

Black holes in the range of two hundred to a million times the mass of the Sun may soon be ruled out (or detected) as a dark-matter candidate by another effect. If such a black hole was to move slowly through a dense cloud, it would produce an intense source of radiation as matter in the cloud fell into the black hole. Several such sources should be observed in our galaxy. A few strange sources of radiation exist in our galaxy that could possibly be black holes of this type, but it is not yet possible to say whether or not these sources are massive black holes.

Meanwhile, efforts to establish the mass of the neutrino have received a spectacular and unexpected assist. For the first time in this century, in early 1987, a supernova occurred in a nearby galaxy, the Large Magellanic Cloud. A burst of neutrinos from this supernova was observed by two neutrino detectors on Earth. The neutrino burst, which presumably was produced when the core of the presupernova star collapsed to form a neutron star, lasted two seconds and contained neutrinos with a wide range of energies.

If neutrinos, like photons, have no mass, then all the neutrinos, regardless of their energies, would have traversed the 150,000 light-years between the Large Magellanic Cloud and Earth at the speed of light. In contrast, if neutrinos have mass, they will travel at less than the speed of light. Low-energy neutrinos would travel more slowly than high-energy neutrinos by an amount that depends on the mass of the neutrino. This spread in speeds would produce a subsequent spread in the arrival times of the neutrinos at Earth. The neutrinos from the supernova arrived almost simultaneously at Earth, producing a sharp pulse, even though they were of energies that varied by as much as 500 percent. John Bahcall and Harvard University physicist Sheldon Glashow used the observed sharpness of the neutrino pulse to set a limit on the mass of the neutrino: It must have a mass less than 1/100,000,000 the mass of the proton. This is about twice as small as the best limit set by terrestrial experiments. It is three times too small a mass if neutrinos are to make the mass density of the universe equal to the critical density. A more detailed analysis by Adam Burrows of the University of Arizona, which takes into account the dynamics of the explosion, agrees with this result. Thus neutrinos of the type observed from the supernova—the electron neutrino produced in the decay of a neutron to a proton and an electron—seem to be eliminated as a solution to the mass-density problem. Electron neutrinos were the type of neutrinos produced in the first three minutes of the Big Bang, the well-understood thermonuclear phase when helium nuclei were synthesized. The hot dark-matter supporters have apparently lost their best candidate. However, other types of neutrinos—the muon and tau neutrinos—might possibly have been produced before the thermonuclear phase, and the experimental limits on these types of neutrinos are much less restrictive, so they remain as possible hot dark-matter candidates.

We can also be encouraged by the possibility that cosmions, if they exist in numbers great enough to solve one or more of the dark-matter problems, will be detected in the laboratory in the near future. Two experiments designed to detect axions through their interaction with a strong magnetic field are under way.

Another group of experiments is designed to detect particles such as photinos and gravitinos that are predicted by superstring and other theories. The experiments would use a detector that consists of superconducting tin grains a few microns in diameter. These grains are surrounded by a magnetic field. One property of a superconducting grain is that a magnetic field cannot exist inside it. But if the grain absorbs energy from a cosmion, it will lose its superconducting properties, the surrounding magnetic field will penetrate into the grain, and a signal will be produced on the readout electronics. Such a detector could reasonably hope to detect anywhere from one to a thousand cosmions per day.

Astrophysicists interested in an apparently unrelated problem have become enthusiastic about the possibility of detecting cosmions. They think that cosmions may be the solution to one of the most persistent and perplexing problems of modern astrophysics—that concerning the solar neutrino. Standard theoretical models of the sun predict that nuclear reactions in the core produce neutrinos at a specific rate. But experiments designed to detect these neutrinos have been running for almost two decades and have found only one third as many neutrinos as predicted. Something is clearly wrong with the standard models.

Dozens of solutions have been proposed. One popular one is that the type of neutrinos produced in the Sun—electron neutrinos—are mostly converted into muon neutrinos before they can escape from the Sun. The neutrino

detector on Earth, which is set up to detect electron neutrinos, will not detect the muon neutrinos. The observed
detection rate would then be lower than expected.

Another scenario uses cosmions. Suppose cold dark
matter is in the form of some cosmion that has a mass
approximately equal to that of a proton. Our galaxy would
then be immersed in a sea of these particles. They would
stream through interstellar space and a few would be captured by stars such as the Sun.

Once inside the Sun, these cosmions could have a significant effect on conditions there. They would collide
with a hot particle in the center and pick up energy from
it. They would then move rapidly away from the center
and eventually give up this extra energy to cooler particles
farther from the center. The net effect is to reduce the temperature in the core of the Sun. Calculations by two independent groups indicate that the decrease of temperature
caused by cosmions inside the Sun could lower the neutrino production rate there enough to bring the predictions into agreement with the observations.

This tidy, resourceful solution is not without problems. None of the currently popular cold-dark-matter candidates, such as photinos, fit the requirements needed to
make the scheme work. Still, the theory of cosmions is
rapidly developing, so it is still possible that such a particle exists. If it does exist, then we should know it within a
few years. If the hypothetical cosmions are abundant
enough and interact with normal matter enough to solve
both the dark-matter and solar-neutrino mysteries, it
should be possible to detect them on Earth in the next few
years with a superconducting grain detector of the type
discussed above.

Observations with gamma-ray telescopes will also provide strong constraints on some cosmion candidates. Hans
Bloemen and Joseph Silk of the University of California at

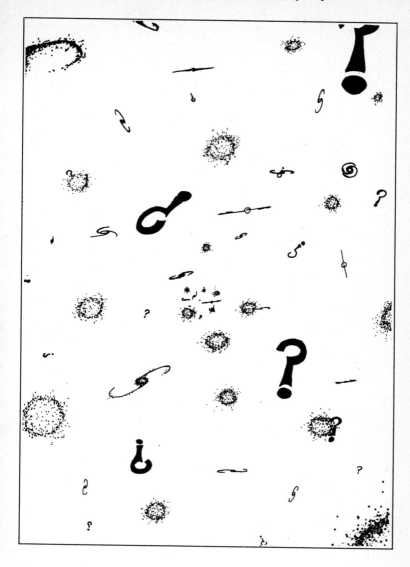

Berkeley have shown that photinos and related relic particles from the very early universe must constitute less than
15 percent of the mass of the bulge and disk of our galaxy.
Otherwise they would produce more gamma rays than are
observed. This limit does not apply directly to the envelope of our galaxy, but is an indirect constraint. Models
for the formation of galaxies from cold dark matter such as
photinos must explain why the bulge and disk of the galaxy contain so few cold cosmions.

The next decade should see a considerable thinning of
the ranks of dark-matter candidates as new observations
and calculations further limit the possibilities. It may
even see the resolution of the problem if, for example, axions of the right number and mass are detected. In the
meantime, the table makes it clear that the solution is not
yet at hand. Also, we cannot ignore the possibility that the
solution may not be listed in the table. Whatever the solution to the mystery of the dark matter, it seems clear that it
will cause a revolution in our understanding of the universe.

CHAPTER
17

The Dark-Matter Mystery: A State of Crisis

The mystery of dark matter in the universe has inexorably drawn the astrophysical community into a state of deepening crisis. They are confronted with a universe where, in the words of Gilbert and Sullivan, "Things are seldom what they seem,/Skim milk masquerades as cream." The evidence is very strong that most of the matter in the universe is in a strange form that no one has ever seen. Observation and theory are often in conflict. Galaxies are not like we thought they were, they did not form in the way we thought they did, our model for the universe may be wrong. Even Einstein's theory of gravity has been questioned. New ideas proliferate like weeds. Which ideas should be kept and which ones discarded? How will they change our perception of the universe?

Over the past fifty years, astrophysicists have been developing a picture as to how the universe works. According to this picture, or paradigm, the universe began with a Big Bang, which produced baryonic matter, electrons, photons, and massless neutrinos. Following the Newton-Einstein law of gravity, virtually all this matter collapsed to form galaxies filled with stars and gas. It was also believed that most of the matter in the galaxies was in some form that could be detected by a radio, infrared, optical, or X-ray telescope. The stars were thought to span a wide range of types, from small red dwarfs about a tenth as massive as the Sun to colossal blue supergiants perhaps a hundred times as massive as the Sun. The distribution of the different types of stars was also thought to be fairly well understood.

The evidence supporting the existence of dark matter shows clearly that something is wrong with this picture. Perhaps many things are wrong. The present picture cannot explain the motions of stars in our galaxy, the rotation rate of spiral galaxies, the amount of hot gas in elliptical galaxies and clusters of galaxies, the motions of galaxies in clusters of galaxies and the Local Supercluster, or the formation of galaxies, clusters, superclusters, and voids.

These problems, many of which started out as curiosities with the work of Oort and Zwicky over fifty years ago, are now serious anomalies that occupy the attention of a large number of astrophyicists. They could signal the overthrow of the existing paradigm. Wolfgang Pauli, a theoretical physicist who played an important role in the quantum revolution of physics in the 1920s and 1930s, must have felt the way many astrophysicists feel today when he said, "In any case, it is too difficult for me, and I wish I had been a movie comedian or something of the sort and had never heard of physics."

If times of crisis are the worst of times for scientists, they are also the best of times. The crumbling of the paradigm liberates the community from the burden of orthodoxy. Wild ideas grow rampantly and creativity flourishes in an invigorating atmosphere that encourages new thoughts, new techniques, and attracts new people to the field. It was in this atmosphere that Pauli, the would-be comedian, was able to postulate the existence of a bizarre new particle, the neutrino, that solved a major problem in quantum physics.

This is the state of astrophysics today. The crisis brought on by the dark-matter mystery causes frustration and discouragement because it is clear that if it cannot be solved, then neither can a host of other important, related problems. It is as if nature is trying to tell us something very obvious and we can't understand. At the same time the crisis causes exhilaration, because astrophysics is alive with new people, such as the elementary-particle physicists, and new ideas, such as the Inflationary Universe, cosmions, cosmic strings, explosive galaxy formation, and the modification of the laws of physics. Experiments are being designed to detect hypothetical particles unimagined a decade ago. If they succeed, a new branch of astronomy will have been born.

Past scientific revolutions have shown us that we should not expect the solution to come like a bolt of lightning. Rather, it will come more like a change of season. The old theories will die out and the correct theories will flourish until they have widespread acceptance in the scientific community.

We have yet to reach that stage in the dark-matter mystery. None of the proposed solutions has won more than a few converts, and the feeling is widespread that something crucial is missing in all of the explanations. One thing that has emerged is that the change in our picture of the universe will be profound, whatever the explanation.

If the dark matter turns out to be in a purely baryonic form, such as brown dwarfs, then we will have to revise our ideas as to how stars and galaxies form. We would have to conclude that for some reason—either natural or perhaps the result of intelligence—tens of trillions of brown dwarfs or Jupiter-like planets must form in a galaxy such as the Milky Way. They must also form in such a way as to leave large dark envelopes around galaxies. We would also have to abandon or radically revise the inflationary Big Bang model for the universe, because in its present form this model requires a higher mass density than is possible in a purely baryonic universe. But if we drop the inflationary Big Bang model, then we drop the simplest explanation for the uniformity of the microwave background radiation. Clearly, major revisions of the theory of the origin of the universe will be necessary if the dark matter is purely baryonic.

The inflationary Big Bang model might be saved if the dark matter is in the form of hot or cold cosmions. Suppose the dark matter in galactic envelopes turns out to be hot cosmions, such as massive neutrinos. Then we have to accept that two types of dark matter exist, because hot cosmions cannot explain the dark matter in the disk of our galaxy. This matter must be something else, perhaps brown dwarfs—a revision of the theory of star formation is then in order. We also have to accept the fact that most of the mass in the universe is in the form of a particle that has never been observed. Envelopes of hot cosmions around galaxies would demand a new method of galaxy formation.

Several interesting proposals have been broached—cosmic strings, explosive galaxy formation, and mock gravity—but they all necessitate a further revision of our present picture. Explosive galaxy formation and mock gravity require the formation of stars before galaxies, which is the reverse of the accepted belief. Cosmic strings are so weird—a diameter smaller than that of a quark and a length of thousands or perhaps even millions of light-years—as to be almost unbelievable. Yet many astrophysicists seriously entertain the possibility that they play a crucial role in the formation of galaxies and clusters of galaxies.

If the dark matter in galactic envelopes is cold cosmions, then revisions of the same order are required. Brown dwarfs or something similar are still needed for the dark matter in the galactic disk. A new type of particle such as a photino is required, and cosmic strings or some new mode of galaxy formation is necessary.

Another possibility is that the solution does not lie in the exotic world of superstring or unified field theories but in a modification of the tried and (believed by most scientists to be) true laws of gravity of Newton and Einstein. Such a modification, which would apply only to very small accelerations, or correspondingly, very large distances, would have profound consequences. All of cosmology would have to be reworked, including our ideas as to how bodies move through the cosmos; they would feel the tug of distant bodies much more strongly than the standard theory predicts. This change would modify the results of calculations of the expansion of the universe—it would take far less matter to slow or stop the expansion of the universe, for example—and thus would radically alter our understanding of the past and future evolution of the universe.

These are a few of the consequences of some of the currently popular proposed solutions to the dark-matter mystery. Other proposals will be brought forward, but it is doubtful that they will be less radical. Once the dark-matter revolution is over, the universe will look different to us than it did before. Either the basic laws of physics may have to be revised, or we may have to accommodate ourselves to a universe in which most of the matter is concealed from us in some dark form such as black holes, brown dwarfs, neutrinos, axions, or photinos. Whatever the resolution, we will understand more clearly how stars, galaxies, and clusters of galaxies form. We should also know more about the origin and ultimate fate of the universe. Will the inflationary Big Bang model of the universe survive this severe test? Or will some alternative model such as a universe with a coasting phase, or a universe that has expanded from a cold rather than a hot state become favored? A cold expanding universe would support many of the unconventional theories of galaxy formation, but it would have difficulty explaining the microwave background radiation and the synthesis of helium.

Scientific revolutions have changed the way people look at the world. Copernicus displaced man from the center of the universe. Newton showed that rational laws could explain the motions of planets. Einstein showed that time is not absolute. The quantum physicists showed that knowledge is probable, not certain. The mystery of the dark matter may be a harbinger of another scientific revolution. Until it is solved, we cannot know for sure. What is certain is that the dark-matter mystery has already spurred the development of new observational techniques and it has opened our imaginations to a myriad of provocative ideas as to the nature of the universe.

Bibliography

We present below some general and technical references for the reader who wishes to know more about a particular subject. *The Hidden Universe* by M. Disney (New York: Macmillan, 1985) is a good general reference to the status of the dark-matter mystery prior to 1984. The basic technical reference is J. Kormendy, ed. *Dark Matter in the Universe* (Dordrecht: D. Reidel, 1987), which is the proceedings of the International Astronomical Union Symposium on Dark Matter in the Universe, held at Princeton University in June 1985. Unless noted below, the quotations in the text are from this symposium or from personal interviews conducted by us. An additional technical reference is V. Trimble, "The Existence and Nature of Dark Matter in the Universe," *Annual Rev, Astron. and Astrophys.*, 25 (1987).

Up-to-date reports on new discoveries relating to the dark-matter mystery can be found in the following nontechnical magazines: *Astronomy, Mercury, Scientific American,* and *Sky and Telescope.*

Chapter 1
Dark Matter

The quotations from Bertrand Russell and John Locke are from B. Russell, *A History of Western Philosophy* (New York, Simon and Schuster, 1945), pp 536, 625.

Chapter 2
The Galactic Disk:
First Evidence for Dark Matter

A reliable guide to the Milky Way galaxy is B. Bok and P. Bok, *The Milky Way* (Cambridge: Harvard Univ. Press, 1981). Jan Oort and his work are described in H. van Woerden, W. Brouw, and H. van de Hulst, eds., *Oort and the Universe* (Dordrecht: D. Reidel, 1980).

Oort's original paper on dark matter in the disk is J. Oort, "The force exerted by the stellar system in the direction perpendicular to the galactic plane and some related problems," *Bull. Astron. Instit. Neth.*, 6: 246 (1932). His update of that work was published as "Note on the Determination of K_v and on the Mass Density Near the Sun," *Bull. Astron. Instit. Neth.*, 15: 45 (1962). John Bahcall's modern version of this research is described in three papers: "The Distribution of Stars Perpendicular to the Galactic Plane," *Astrophys. J.*, 276: 156 (1984), "Self-Consistent Determination of the Total Amount of Matter Near the Sun," *Astrophys. J.*, 276: 169 (1984), and "K Giants and the Total Amount of Matter Near the Sun," *Astrophys. J.*, 287: 926 (1984). See also Bahcall's article, "Dark Matter in the Galactic Disk" in Kormendy (1987).

Chapter 3
The Galactic Disk:
Dark-Matter Candidates

White dwarfs, neutron stars, and black holes are discussed in G. Greenstein, *Frozen Star* (New York: Freundlich, 1983), and W. Tucker and R. Giacconi, *The X-ray Universe* (Cambridge: Harvard Univ. Press, 1985). Infrared astronomy and the search for brown dwarfs is discussed in W. Tucker and K. Tucker, *The Cosmic Inquirers* (Cambridge: Harvard Univ. Press, 1986). The synthesis of helium from hydrogen in the early universe is discussed in J. Silk, *The Big Bang* (San Francisco: W. H. Freeman and Co., 1980), R. Wagoner and D. Goldsmith, *Cosmic Horizons* (San Francisco: W. H. Freeman and Co., 1982), and S. Weinberg, *The First Three Minutes* (New York: Bantam, 1979).

The number of low-mass stars in the galaxy is discussed by R. Probst and R. O'Connell in "The Luminosity Function of Very Low Mass Stars," *Astrophys. J.*, 252: L69 (1982) and by G. Gilmore and P. Hewett in *Nature*, 306: 669 (1983). White dwarfs as a possible solution to the dark-matter problem in the galactic disk is discussed in K. Olive, "G Dwarfs, White Dwarfs and the Local Dark Matter

Problem," *Astrophys. J.*, 309: 210 (1986) and references cited therein. The search for brown dwarfs in Infrared Astronomical Satellite data is described in F. Low's review, "IRAS Results on Dark Matter in Our Galaxy" in Kormendy (1987).

The evaporation of hydrogen rocks is discussed by D. Hegyi and K. Olive in "A Case Against Baryons in Galactic Halos," *Astrophys. J.*, 303: 56 (1986). The composition of the oldest stars in the galaxy is discussed in a review by J. Jones in "Early Galactic Evolution and the Nature of the First Stars," *Pub. Astron. Soc. Pacific*, 97: 593 (1985).

Chapter 4
Beyond the Luminous Edge:
A Galactic Envelope of Dark Matter

A good general reference is B. Bok and P. Bok, *The Milky Way* (Cambridge: Harvard Univ. Press, 1981), which discusses research on the rotation of the Milky Way galaxy from Oort to the present day. An overview of the new results, with special reference to the dark halo around our galaxy, is given in B. Bok, "The Milky Way Galaxy," *Scientific American*, March 1981, and V. Rubin, "Dark Matter in Spiral Galaxies," *Scientific American*, June 1983. Both these articles are in P. Hodge, ed., *The Universe of Galaxies* (New York: W. H. Freeman and Co., 1984).

More technical reviews that give references to the work discussed in this chapter are L. Blitz, M. Fich, and S. Kulkarni, "The New Milky Way," *Science*, 22: 1233 (1983), and V. Rubin, "The Rotation of Spiral Galaxies," *Science*, 220: 1339 (1983b). See also B. Carney and D. Latham, "Escape Velocity from the Galaxy," in Kormendy (1987), and E. Olszewski, R. Peterson, and M. Aaronson, "High Precision Radial Velocities for Faint Giants: Radial Velocities of Extreme Halo Systems and the Mass of the Galaxy," *Astrophys. J.*, 302: L45–48 (1986).

Chapter 5
Cosmions: Hot and Cold

The production of neutrinos in the hot Big Bang model for the universe is discussed in J. Silk, *The Big Bang* (San Francisco: W. H. Freeman and Co. (1980), R. Wagoner and D. Goldsmith, *Cosmic Horizons* (San Francisco: W. H. Freeman and Co., 1982), S. Weinberg, *The First Three Minutes* (New York: Bantam, 1979), H. Pagels,

Perfect Symmetry (New York: Simon and Schuster, 1985), and J. Trefil, *The Moment of Creation* (New York: Macmillan, 1984) describe theories of elementary particles and their impact on astrophysics and cosmology.

A review of the advantages and disadvantages of different types of cosmions in forming galaxies is given in G. Blumenthal, S. Faber, J. Primack, and M. Rees, "Formation of Galaxies and Large-Scale Structure with Cold Dark Matter," *Nature*, 311: 517 (1984).

Chapter 6
The Turning Point:
Dark Matter in Spiral Galaxies

Descriptions of the research on dark matter in spiral galaxies are given in V. Rubin, "Dark Matter in Spiral Galaxies," *Scientific American*, June 1983, and "The Rotation of Spiral Galaxies," *Science*, 220: 1339 (1983).

On a more technical level, see also the excellent and influential review of dark matter by S. Faber and J. Gallagher, "Masses and Mass-to-Light Ratios of Galaxies," *Ann. Rev. Astr. Astrophys.*, 17: 135 (1979), as well as V. Rubin, "Constraints on Dark Matter Properties from Optical Rotation Curves" and T. S. van Albada and R. Sancisi, "HI Rotation Curves," both in Kormendy (1987).

Chapter 7
Dwarf Galaxies

Zwicky's ideas concerning the inexhaustibility of nature and the method of negation and subsequent construction are presented in F. Zwicky, *Morphological Astronomy* (Berlin: Springer-Verlag, 1957) and F. Zwicky, *Catalog of Selected Compact Galaxies and of Post-Eruptive Galaxies* (Zurich: L. Spreich, 1971). An old but useful general review of dwarf galaxies is given by P. Hodge in "Dwarf Galaxies," *Ann. Rev. Astron. Astrophys.*, 9: 35 (1971).

Marc Aaronson's work is described in "Accurate Radial Velocities for Carbon Stars in Draco and Ursa Minor: The First Hint of a Dwarf Spheroidal Mass-to-Light Ratio," *Astrophys. J.*, 266: L11 (1983) and in "The Search for Dark Matter in Draco and Ursa Minor: A Three Year Progress Report" in Kormendy (1987). Dark matter in dwarf galaxies is also discussed in the same volume by S. Faber and D. Lin in "Is There Nonluminous Matter in Dwarf Spheroidal Galaxies?," p. L17, and "Some Implications of Nonluminous

Matter in Dwarf Spheroidal Galaxies," P. L21. A dissenting view is presented by J. Cohen in "The Velocity Dispersion of the Globular Clusters in the Fornax Dwarf Galaxy," *Astrophys.J.*, 270: L41 (1983). The energy state, or phase space constraint that rules out neutrinos as dark-matter candidates for dwarf galaxies was first discussed by S. Tremaine and J. Gunn in "Dynamical Role of Light Neutral Leptons in Cosmology," *Physical Review Letters*, 42: 407 (1979). The work of Frogel and Seitzer is presented in J. Frogel and P. Seitzer, "Radial Velocities of Carbon Stars in Three Dwarf Spheroidal Galaxies," *Astronom. J.*, 90: 1796 (1985). The paper by Lake and Schommer is "Mass-to-Light Ratios for Binary Pairs of Dwarf Irregular Galaxies," *Astrophys. J.*, 279: L19 (1984).

Chapter 8
Hot Gas and Dark Matter

A nontechnical discussion of the development of X-ray astronomy is discussed in W. Tucker and R. Giacconi, *The X-ray Universe* (Cambridge: Harvard Univ. Press, 1985).

The heredity-versus-environment argument as to the origin of elliptical galaxies is reviewed by A. Dressler in "The Evolution of Galaxies in Clusters," *Ann. Rev. Astron. Astrophys.*, 22: 185 (1984). X-ray evidence for dark matter around elliptical galaxies is given by W. Forman, C. Jones, and W. Tucker in "Hot Coronae Around Early-Type Galaxies," *Astrophys. J.*, 293: 102 (1985), and in C. Canizares's contribution, "X-ray Halos in Galaxies and Clusters: Theory" in Kormendy (1987).

Chapter 9
Gravitational Lenses:
Detecting Dark Matter with Bent Light

A popular account of Einstein's calculations of the bending of light by a gravitational field is given in N. Calder, *Einstein's Universe* (New York: Penguin, 1980). Nontechnical treatments of gravitational lenses are given by J. Lawrence in "Gravitational Lenses and the Double Quasars," *Mercury*, 9: 66 (1980) and by F. Chaffee, Jr., in "The Discovery of a Gravitational Lens," *Scientific American* November 1983, p. 70.

Einstein's original paper with the mistake is in his paper "On the Influence of Gravitation on the Propagation of Light," *Annalen der Physik*, 35 (1911). The mistake is corrected in his paper "Foundation of the General Theory of Relativity," *Annalen der Physick,* 49 (1916). Both papers have been reprinted in the anthology Albert Einstein et al., *The Principle of Relativity* (New York: Dover, 1924).

The work of Tyson's group is described in J. A. Tyson et al., "Galaxy Mass Distribution from Gravitational Light Deflection," *Astrophys. J.,* 281: L59 (1984) and in J. Tyson, "Image Distortion Near Galaxies: Dwarfs or Lensing?," *Nature,* 316: 799 (1985).

Chapter 10
Dark Matter in Groups and Clusters of Galaxies

Clusters of galaxies and X-ray observations of them are discussed in a nontechnical way in W. Tucker and R. Giacconi, *The X-ray Universe* (Cambridge: Harvard Univ. Press, 1985), W. Tucker and K. Tucker, *The Cosmic Inquirers* (Cambridge: Harvard Univ. Press, 1986), and in P. Gorenstein and W. Tucker, "Rich Clusters of Galaxies," *Scientific American,* November 1978. Interactons between galaxies is discussed by S. Strom and K. Strom in "The Evolution of Galaxies," *Scientific American,* April 1979, and by A. Toomre and J. Toomre in "Violent Tides Between Galaxies," *Scientific American,* December 1973. Both these articles are reproduced in P. Hodge, ed., *The Universe of Galaxies* (New York: W.H. Freeman and Co., 1984).

Zwicky's first paper on dark matter in the Coma clusters of galaxies was in *Helv. physics acta,* 6: 110 (1933). More accessible and more detailed is F. Zwicky, "On the Masses of Nebulae and Clusters of Nebulae," *Astrophys. J.,* 86: 217 (1937). Similar results for the Virgo cluster were reported by S. Smith in "The Mass of the Virgo Cluster," *Astrophys. J.,* 83: 499 (1936). The mass determination of the Virgo cluster by X-ray observatons is discussed by D. Fabricant and P. Gorenstein in "Further Evidence for M87's Massive Dark Halo," *Astrophys. J.,* 267: 535 (1983). The *Einstein* X-ray Observatory results on clusters of galaxies are reviewed by W. Formen and C. Jones, "X-ray Imaging Observations of Clusters of Galaxies," *Ann. Rev. Astron. Astrophys,* 20: 547 (1982).

Early discussions of dark matter in groups of galaxies are given in the paper by J. Ostriker, J. Peebles, and A. Yahil, "The Size and Mass of Galaxies and the Mass of the Universe," *Astrophys. J.,* 193: L1 (1974) and in the paper by J. Einasto, A. Kaasik, and E. Saar in *Nature,* 250: 309 (1974). Burbidge's critique is in "On the Masses and Relative Velocities of Galaxies," *Astrophys. J.,* 196: L7 (1975). S. Faber and J. Gallagher, "Masses and Mass-to-Light Ratios of Galaxies," *Ann. Rev. Astro. Astrophys.*, 17: 135 (1979) give a critical review of both sides of the issue. An update is given in Ostriker, "Masses from Satellites of Galaxies" and M. Davis, "Masses of Groups and Clusters of Galaxies," both in Kormendy (1987).

Byrd and Valtonen's work on galaxy groups is described in "Origin of Redshift Differentials in Galaxy Groups," *Astrophys. J.*, 289: 535 (1985). This work uses the catalog compiled by Huchra and Geller, which appears in "Groups of Galaxies. I. Nearby Groups," *Astrophys. J.*, 257: 423 (1982). The skeptical view of clusters is M. Valtonen, K. Innanen, T. Huang, S. Saarinen, "No missing mass in clusters of galaxies?," *Astron. Astrophys.*, 143: 182 (1985). The X-ray observations relating to the evolution of clusters of galaxies are described in Forman and Jones (1982).

Chapter 11
The Search for Dark Matter
in the Local Supercluster

Nontechnical accounts of modern research on superclusters are N. Morrison and D. Morrison, "The Local Supercluster and the Large Scale Structure of the Universe," *Mercury*, September 1982, and S. Gregory and L. Thompson, "Superclusters and Voids in the Distribution of Galaxies," *Scientific American*, March 1982.

More technical are the reviews by J. Peebles, "The Origin of Galaxies and Clusters of Galaxies," *Science*, 224: 1385 (1984), J. Oort, "Superclusters," *Ann, Rev. Astr. Astrophys.*, 21: 373 (1983), and M. Davis and J. Peebles, "Evidence for Local Anisotropy of the Hubble Flow," Ibid., 109. The term *Local Supercluster* appears to have been coined by de Vaucouleurs in an article in *Astron. J.*, 58: 30 (1953). The definitive modern work is R. Tully, "The Local Supercluster," *Astrophys. J.*, 257: 389 (1982). The work of Aaronson and colleagues is discussed in M. Aaronson, G. Bothun, J. Mould, J. Huchra, R. Schommer, and M. Cornell, "A Distance Scale from the Infrared Magnitude/HI Velocity-Width Relation: V. Distance Moduli to 10 Galaxy Clusters and Positive Detection of Bulk Supercluster Motion Toward the Microwave Anisotropy," *Astrophys. J.*, 302: 536 (1986). The uncertainties in the estimate of dark matter in the Local Supercluster are discussed by H. Lee, Y. Hoffman, and C. Ftaclas in "On the Application of the Spherical Infall Model to the Local Supercluster," *Astrophys. J.*, 304: L11, and by J. Williams and M. Davis in "Velocity Fields Around Rich Clusters of Galaxies," *Astrophys.J.*, 308: 499.

Chapter 12
The Search for a
Universal Sea of Dark Matter

A nontechnical presentation of the issue of whether the universe is finite or infinite is given in J. Gott III, J. Gunn, D. Schramm, B. Tinsley, "Will the Universe Expand Forever?," *Scientific American*, March 1976 (reprinted in *Cosmology + 1* (San Francisco: W. H. Freeman and Co., 1977); see also J. Silk, *The Big Bang* (San Francisco: W. H. Freeman and Co., 1980) and R. Wagoner and D. Goldsmith, *Cosmic Horizons* (San Francisco: W. H. Freeman and Co., 1982).

More technical discussions are given by J. Peebles in "The Mean Mass Density of the Universe," *Nature*, 321: 27 (1986), A. Boesgaard and G. Steigman in "Big Bang Nucleosynthesis: Theories and Observations," *Ann. Rev. Astr. Astrophys.*, 23: 319 (1985), in J. Audouze's contribution to Kormendy (1987), and by J. Huchra in "On the Determination of Cosmological Parameters" in *Innerspace/Outerspace*, eds. E. Kolb et al. (Univ. Chicago Press, Chicago, 1986).

The first results of the Center for Astrophysics survey are discussed in V. de Lapparent, M. Geller, and J. Huchra, "A Slice of the Universe," *Astrophys. J.*, 302: L1, (1986). Large-scale streaming motions in the local universe are discussed in A. Dressler et al., "Spectroscopy and Photometry of Elliptical Galaxies: A Large Scale Streaming Motion in the Local Universe," *Astrophys. J.*, 313: L37 (1987). The discovery of galaxies in the Boötes void is reported in J. Moody et al., "Emission line galaxies in the Boötes void," *Astrophys. J.*, 314: L33 (1987).

The results from the latest use of the red-shift method are given by E. Loh and E. Spillar in "A Measurement of the Mass Density of the Universe," *Astrophys. J.*, 307: L1 (1986).

Chapter 13
Dark Matter and the
Inflationary Universe

Nontechnical discussions of the difficulties with the standard Big Bang model are given in A. Guth and P. Steinhardt, "The Inflationary Universe," *Scientific American*, May 1984, A. Linde, "The Universe: Inflation out of Chaos," *New Scientist*, March 7, 1985, p. 14, and in H. Pagels, *Perfect Symmetry* (New York: Simon and Schuster, 1985) and J. Trefil, *The Moment of Creation* (New York: Macmillan, 1984).

Kazanas's and Guth's papers on the Inflationary Universe are D. Kazanas, "Dynamics of the Universe and Spontaneous Symmetry Breaking," *Astrophys. J.*, 241: L59 (1980) and A. Guth, "The Inflationary Universe: A Possible Solution to the Horizon and Flatness Problems," *Physical Review D*, 23: 347 (1980).

Chapter 14
Dark Matter and Model Universes

The standard model for galaxy formation is discussed in a nontechnical way in J. Silk, *The Big Bang* (San Francisco: W. H. Freeman and Co., 1980). Nontechnical discussions of Grand Unified Theories, and supersymmetric theories and their relevance to cosmology are given in H. Pagels, *Perfect Symmetry* (New York: Simon and Schuster, 1985) and J. Trefil, *The Moment of Creation* (New York: Macmillan, 1984). Cosmic strings are discussed by P. Davies in "Relics of Creation," *Sky and Telescope*, February 1985, p. 112.

Technical treatments are given in J. Gott, "Recent Theories of Galaxy Formation," *Ann. Rev. Astron. Astrophys.*, 15: 235 (1977), J. Peebles, "The Origin of Galaxies and Clusters of Galaxies, *Science*, 224: 1385 (1984), G. Blumenthal et al., "Formation of Galaxies and Large-Scale Structure with Cold Dark Matter," *Nature*, 311: 517 (1984). The explosive-galaxy-formation model is discussed in J. Ostriker and L. Cowie, "Galaxy Formation in an Intergalactic Medium Dominated by Explosions," *Astrophys. J.*, 243: L127 (1981), E. Vishnaic, J. Ostriker, and E. Bertschinger, "Explosions in the Early Universe," *Astrophys. J.*, 291: 399 (1985), and A. Wandel, "Cooling Shells and Galaxy Formation in the Early Universe," *Astrophys. J.*, 294: 385 (1985), and S. Ikeuchi, *Pub. Astron. Soc. Japan*, 33: 211 (1981).

The basic references for model universe calculations are White's contribution to Kormendy (1987), S. White, C. Frenk, and M. Davis, "Clustering in a Neutrino-Dominated Universe," *Astrophys. J.*, 274: L1–L6 (1983) and M. Davis, G. Efstathiou, C. Frenk, and S. White, "The Evolution of Large-Scale Structure in a Universe Dominated by Cold Dark Matter," *Astrophys. J.*, 292: 371–394 (1985). An application to the formation of galaxies is given in C. Frenk, S. White, G. Efstathiou, and M. Davis, "Cold Dark Matter, the Structure of Galactic Haloes and the Origin of the Hubble Sequence," *Nature*, 317: 595–597 (1985). The analytical approach is discussed in Blumenthal et al. (1984), in G. Blumenthal, S. Faber, R. Flores, and J. Primack, "Contraction of Dark Matter Halos Due to Baryonic Infall," *Astrophys. J.* 301, 27 (1986), and in J. Primack, "Dark Matter, Galaxies, and Large Scale Structure in the Universe," lectures presented at the International School of Physics, "Enrico Fermi," Varenna, Italy (1984). The difficulties posed by the smoothness of the microwave background radiation are discussed in J. Bond and G. Efstathiou, "Cosmic Background Radiation Anisotropies in Universes Dominated by Nonbaryonic Dark Matter," *Astrophys. J.*, 285: L44–L48 (1984), N. Vittorio and J. Silk, "Fine-Scale Anisotrophy of the Cosmic Microwave Background in a Universe Dominated by Cold Dark Matter," ibid., L39–L43, and in N. Vittorio and J. Silk, "Scale-Invariant Density Perturbations, Anisotropy of the Cosmic Microwave Background and Large-Scale Peculiar Velocity Field," *Astrophys. J.*, 293: L1 (1985). The effects of reheating the universe on the microwave background and galaxy formation are discussed by J. Peebles in "Cosmic Background Temperature Anisotropy in a Minimal Isocurvature Model for Galaxy Formation," *Astrophys. J.*, 315: L73 (1987).

A comprehensive review of biased galaxy formation is given by A. Dekel and M. Rees in "Physical mechanisms for biased galaxy formation," *Nature*, 326: 452 (1987). The effect on model universes is discussed in S. White et al., "Clusters, Filaments and Voids in a Universe Dominated by Cold Dark Matter," *Astrophys. J.*, 313: 505 (1987).

Technical treatments of the role of cosmic strings in galaxy formation are given by A. Vilenkin, in *Phys. Rep.*, 121: 263 (1985), C. Hogan and M. Rees in *Nature*, 311: 109 (1984), and N. Turok, *Phs. Rev. Lett*, 55: 1801 (1985). Mock gravity is discussed by C. Hogan and S. White in *Nature*, 321: 575 (1986). The role of vibrating cosmic strings in the formation of galaxies and voids is discussed in J. Ostriker, C. Thompson, and E. Witten, "Cosmological Effects of Superconducting Strings," *Phys. Lett. B.* 180: 231 (1986).

Chapter 15
Dark Matter or a New Law of Gravity?

The basic references to Milgrom's work are his contribution to Kormendy (1987), M. Milgrom, "A Modification of the Newtonian Dynamics as a Possible Alternative to the Hidden Mass Hypothesis"; "A Modification of the Newtonian Dynamics: Implications for Galaxies"; "A Modification of Newtonian Dynamics: Implications for Galactic Systems," in *Astrophys. J.*, 270: 365, 371, 384 (1985), and J. Beckenstein and M. Milgrom, "Does the Missing Mass Problem Signal the Breakdown of Newtonian Gravity?," *Astrophys. J.*, 286: 7 (1984). Problems with the modified dynamics are discussed by J. Felten in "Milgrom's Revision of Newton's Laws: Dynamical and Cosmological Consequences," ibid, 3, and J. Primack, "Dark Matter, Galaxies, and Large Scale Structure in the Universe," lectures presented at the International School of Physics, "Enrico Fermi," Varenna, Italy (1984), L. Hernquist and P. Quinn, "Shell Galaxies and Alternatives to the Dark Matter Hypothesis," *Astrophys. J.*, 312: 17 (1987), and in R. H. Sanders, "Alternatives to Dark Matter," European Southern Observatory Preprint #459, August 1986. Sanders's modification of Milgrom's hypothesis to include antigravity is given in Sanders (1986).

Chapter 16
An Evaluation of the
Dark-Matter Mystery

Shadow matter is discussed by E. Kolb, D. Seckel, and M. Turner in "The Shadow World of Superstring Theories," *Nature*, 314: 415 (1985), and L. Kraus et al. in "Inflation and Shadow Matter," *Nature*, 319: 748 (1986).

Artificial brown dwarfs and the explanation for dark matter is discussed by D. Criswell in "Solar System Industrialization: Implications for Interstellar Migration" in eds., B. Finney and E. Jones, *Interstellar Migration and the Human Experience* (Berkeley: Univ. California Press, 1985).

The ratings given here are our own, but for the most part they agree with the consensus of the astrophysical community, as reflected in the following reviews: Martin Rees and J. Gunn's contribution to Kormendy (1987); B. J. Carr, "Black Holes, Pregalactic Systems and the Dark Matter Problem," lectures given at Santander Summer School on Relativistic Astrophysics and Cosmology, September 3–7, 1984; J. Primack, "Dark Matter, Galaxies, and Large Scale Structure in the Universe," lectures presented at the International School of Physics, "Enrico Fermi" Varenna, Italy (1984); and D. Schramm, "Phase Transitions and Dark Matter Problems" in *Nuclear Physics*, B 252: 53 (1985).

The limit on the neutrino mass derived from supernova observations is reported by J. Bahcall and S. Glashow in "Upper limit on the mass of the electron neutrino," *Nature*, 326: 476 (1987).

Attempts to detect cosmions are discussed in J. Moody's contribution to Kormendy (1987), in P. Sikivie, *Phys. Rev. Lett.*, 51: 1415 (1983), in A. Drukier, G. Gelmini, and S. Murray, "Detecting Solar Cosmions (submitted to *Science 1987*), and in L. Krauss, "Dark Matter in the Universe," *Scientific American*, November 1986. Indirect effects of cold dark matter are discussed by J. Silk and H. Bloemen in "A Gamma Ray Constraint on the Nature of Dark Matter," *Astrophys. J.*, 313: L47 (1987).

Chapter 17
The Dark-Matter Mystery:
A State of Crisis

The nature of scientific revolutions is discussed in S. Tremaine's contribution to Kormendy (1987), in T. Kuhn, *The Structure of Scientific Revolutions*, 2nd edit. (Chicago: Univ. Chicago Press, 1970), and in I. Cohen, *Revolution in Science* (Cambridge: Harvard Univ. Press, 1985). A cold expanding universe is discussed in D. Layzer, *Constructing the Universe* (New York: W. H. Freeman and Co., 1984).

Index

About the Authors

Wallace Tucker was born in Oklahoma in 1939. He received his Ph.D. in physics from the University of California at San Diego, and has worked in the field of X-ray astronomy for over twenty years. He is currently a visiting professor at the University of California at Irvine, and a research astrophysicist at the Smithsonian Astrophysical Observatory in Cambridge, Massachusetts. He and his wife, Karen, have collaborated on numerous popular-science articles and a popular-astronomy book, *The Cosmic Inquirers*.